SUSTAINABILITY IN ENERGY AND ENVIRONMENT

Engineered Materials and Smart Computational Techniques

SUSTAINABILITY IN ENERGY AND ENVIRONMENT

Engineered Materials and Smart Computational Techniques

Edited by

Ann Rose Abraham, PhD

Heru Susanto, PhD

A. K. Haghi, PhD

Kian Hariri Asli

AAP | APPLE ACADEMIC PRESS

First edition published 2025

Apple Academic Press Inc.
1265 Goldenrod Circle, NE,
Palm Bay, FL 32905 USA

760 Laurentian Drive, Unit 19,
Burlington, ON L7N 0A4, CANADA

CRC Press
2385 NW Executive Center Drive,
Suite 320, Boca Raton FL 33431

4 Park Square, Milton Park,
Abingdon, Oxon, OX14 4RN UK

Library and Archives Canada Cataloguing in Publication

CIP data on file with Canada Library and Archives

Library of Congress Cataloging-in-Publication Data

CIP data on file with US Library of Congress

ISBN: 978-1-77491-620-9 (hbk)
ISBN: 978-1-77491-621-6 (pbk)
ISBN: 978-1-00349-212-2 (ebk)

About the Editors

Ann Rose Abraham, PhD
Assistant Professor, Sacred Heart College (Autonomous), Thevara, Kochi, Ernakulam, Kerala, India

Ann Rose Abraham, PhD, is currently an Assistant Professor at the Department of Physics, Sacred Heart College (Autonomous), Thevara, Kochi, Kerala, India. She has expertise in the fields of condensed matter physics, nanomagnetism, multiferroics, polymeric nanocomposites, etc. She has research experience at various reputed national institutes, including Bose Institute, Kolkata, India; SAHA Institute of Nuclear Physics, Kolkata, India; and UGC-DAE CSR Center, Kolkata, India; and she has collaborated with various international laboratories. She is a recipient of a young researcher award in physics and best paper awards (2020 and 2021). She served as an Assistant Professor and Examiner at the Department of Basic Sciences, Amal Jyothi College of Engineering, under APJ Abdul Kalam Technological University, Kerala, India. Dr. Abraham is a frequent speaker at national and international conferences. She has a good number of publications to her credit in many peer-reviewed high-impact journals of international repute. She has authored many book chapters and edited more than 10 books with Taylor and Francis, Elsevier, etc.

Heru Susanto, PhD
Researcher, Center for Innovative Engineering/School of Business, University of Technology Brunei, Darussalam; Senior Researcher, National Research and Innovation Agency, Indonesia; Honorary Professor, Information Management, Tunghai University, Taiwan

Heru Susanto, PhD, is a Researcher at the Center for Innovative Engineering/School of Business at the University of Technology Brunei. He is also a senior researcher of the Data Security Group at the National Research and Innovation Agency. At the same time, he is also an honorary professor at the Department of Information Management, College of Management, Tunghai University, Taichung, Taiwan. Dr. Susanto has experience as an IT professional and as web division head of IT Strategic Management at Indomobil Group Corporation. He has worked as the Prince Muqrin Chair for Information Security Technologies at King Saud University, Riyadh, Saudi Arabia. His research interests are in the areas of information systems, computer security, information security management systems, and information systems and technology as an enabler of organizations' business process reengineering and e-marketing. He has over 15 years of professional teaching and research experience at the university level and over 20 years of working on research for various studies in the field. He has published more than 140 publications, including seven authored books and 60 book chapters and more than 65 international publications in peer-reviewed journals and proceedings in Scopus, Science Direct, Springer, ISI, etc.

A. K. Haghi, PhD

Professor Emeritus of Engineering Sciences and Senior Material Engineering Consultant, UK

A. K. Haghi, PhD, is a retired Professor with over 250 academic research-oriented books as well as over 1000 research papers published in various journals and conference proceedings with over 3350 citations and an h-index of 30, according to the Google Scholar database. He is currently a senior material engineering consultant in the UK. Professor A. K. Haghi has received several grants, consulted for several major corporations, and is a frequent speaker to national and international audiences. He is the Founder and former Editor-in-Chief of the *International Journal of Chemoinformatics and Chemical Engineering* and *Polymers Research Journal.* Professor A. K. Haghi has acted as an editorial board member of many international journals. He has served as a member of the Canadian Research and Development Center of Sciences and Cultures and the Research Chemistry Center, Coimbra, Portugal. He has supervised several PhD and MSc theses at the University of Guilan (UG) and co-supervised international doctoral projects. Professor Haghi holds a BSc in urban and environmental engineering from the University of North Carolina (USA) and holds two MSc degrees, one in mechanical engineering from North Carolina State University (USA) and another one in applied mechanics, acoustics, and materials from the Université de Technologie de Compiègne (France). He was awarded a PhD in engineering sciences at Université de Franche-Comté (France).

Kian Hariri Asli
Research Scholar, Department of Electronic Engineering,
University of Rome "Tor Vergata," Rome, Italy

Kian Hariri Asli is a research scholar in the Department of Electronic Engineering, University of Rome "Tor Vergata," Rome, Italy. He has expertise in electronic and control engineering. He works toward the development of energy saving by networked sensors, fuzzy logic, artificial intelligence (AI), and the Internet of Things (IoT) based on geospatial information systems (GIS).

Contents

Contributors

Mohammad Ashfaq
Department of Biotechnology and University Center for Research and Development (UCRD), Chandigarh University, Gharaun, Mohali, Punjab, India

Kian Hariri Asli
Department of Electronic Engineering, University of Rome "Tor Vergata," Rome, Italy

Brahmananda Chakraborty
High Pressure and Synchrotron Radiation Physics Division, Bhabha Atomic Research Center, Trombay, Mumbai, Maharashtra, India; Homi Bhabha National Institute, Mumbai, Maharashtra, India

Divya Chauhan
Department of Chemical and Biomedical Engineering, University of South Florida, Tampa, Florida, United States

Arif Junuin
Center for Research Collaboration of Graph Theory and Combinatory, UTB Brunei Darussalam–BRIN Indonesia–THU Taiwan–ITB Indonesia–UI Indonesia, and National Research and Innovation Agency, Jakarta, Indonesia

M. V. Manjusha
Assistant Professor, Department of Physics, The Cochin College, Kochi, Ernakulam, Kerala, India

K. Nayana
Department of Physics, Government Victoria College, Palakkad, Kerala,
Affiliated to University of Calicut, India; Department of Physics, N. S. S. College, Ottapalam, Palakkad, Kerala, Affiliated to University of Calicut, India

S. S. Praveen
Assistant Professor in Physics, St. Thomas College, Kozhencherry, Pathanamthitta, Kerala, India

Dhanya Raj
Research Scholar, Department of Physics, The Cochin College, Kochi, Ernakulam, Kerala, India

Gopal Sanyal
Mechanical Metallurgy Division, Bhabha Atomic Research Center, Trombay, Mumbai, Maharashtra, India

Nidhi Sharma
Department of Physics, Dhanauri P.G. College, Dhanauri, Roorkee Haridwar, Uttarakhand, India

A. P. Sunitha
Department of Physics, Government Victoria College, Palakkad, Kerala,
Affiliated to University of Calicut, India

Alifya Kaya Shafa
Department of Computer Science, Tunghai University, Taichung City, Taiwan

Heru Susanto
Center for Innovative Engineering and School of Business, Universiti Teknologi Brunei, Mukim Gadong, Brunei

Neetu Talreja
Department of Science, Faculty of Science and Technology, Alliance University, Anekal, Bangalore, Karnataka, India

Ravi Kumar Trivedi
High Pressure and Synchrotron Radiation Physics Division, Bhabha Atomic Research Center, Trombay, Mumbai, Maharashtra, India

T. V. Venkidesh
Research Scholar, Department of Physics, The Cochin College, Kochi, Ernakulam, Kerala, India

Aswathy Vijayan
Doctoral Student, Department of Physics, St. Thomas College, Kozhencherry, Pathanamthitta, Kerala, India

Preface

This research-oriented book covers integrated, sustainable design for energy and environmental engineering and computer science. In this new title, the key components of energy are described, and the main elements of sustainability are introduced with an emphasis on the sustainable development of the environment. The volume is broadly divided into the themes of energy, environment, and sustainable development. It also covers smart computational technologies in the domain of energy and environmental systems engineering.

With the formal definition of sustainability, this volume offers seven chapters that cover topics such as the fundamentals, demand, management, technology, and analytical models in two advanced sections, bringing together both fundamental aspects and practical knowledge from energy engineering, environmental engineering, and computer science. In the first section of this book, components of energy sustainability are described and examined with various other factors. To enhance energy sustainability and sustainable development, different case studies with a focus on environmental perspectives are presented in the second section.

This book will be a valuable reference book for professionals and research students interested in the different fields of energy, environment, and sustainable development.

PART I
Innovative Sustainable Energy Engineering

CHAPTER 1

Electrochemical Energy Storage in 2D Nanomaterials: Present Status and Future Challenges

GOPAL SANYAL,[1] RAVI KUMAR TRIVEDI,[2] and BRAHMANANDA CHAKRABORTY[2,3]

[1]*Mechanical Metallurgy Division, Bhabha Atomic Research Center, Trombay, Mumbai, Maharashtra, India*

[2]*High Pressure and Synchrotron Radiation Physics Division, Bhabha Atomic Research Center, Trombay, Mumbai, Maharashtra, India*

[3]*Homi Bhabha National Institute, Mumbai, Maharashtra, India*

ABSTRACT

Supercapacitors (SCs) are widely recognized as a preferred energy storage device for electricity. They have emerged as a viable alternative to traditional capacitors and fuel cells due to their higher energy density and power concentration compared to batteries. However, they do have some limitations such as lower energy density and shorter lifespan compared to batteries. To overcome these challenges, scientists are working on creating 2D nanostructures and modifying them using various techniques like nobbling, introducing oxygen vacancles, and combining them with carbon allotropes to enhance their electrochemical properties. In this chapter, we have presented the aforementioned themes with the assistance of newly published reports. Furthermore, we have provided theoretical insights based on density functional theory (DFT) for the electrochemical response of the probe substances, as reported in recent literature. This assessment succinctly

Sustainability in Energy and Environment: Engineered Materials and Smart Computational Techniques. Ann Rose Abraham, Heru Susanto, A. K. Haghi, & Kian Hariri Asli (Eds.)

highlights the progress in the field of supercapacitor energy storage and the various methods involved in the fabrication of supercapacitor electrode materials, which will be highly beneficial to researchers working in the field of energy storage. Additionally, the challenges and future perspectives of this exciting research endeavor are thoroughly discussed.

1.1 INTRODUCTION

The current global electricity production capacity is around 20 terawatt hours [1]. Figure 1.1 shows how electricity is generated from different sources [2]. Despite efforts to preserve and develop new skills, experts predict that the global electricity demand will double by the middle of this century and triple by the end of the century. Electricity is the dominant form of energy consumption, and the demand for electricity is growing at a faster pace than overall energy consumption across all sources. At the same time, natural gas and oil extraction are projected to increase in the coming years. Coal has been the leading source of electricity production [3]; abundant coal resources may sustain current consumption levels compared to oil and gas. However, every kWh of electricity produced by burning coal generates an average of 1 kg of lifecycle CO_2 emissions, a greenhouse gas that is widely considered the main contributor to global warming [4, 5]. In the USA alone, coal power generation units emit 1.5 billion tons of CO_2 annually, and emissions from developing nations are increasing. Many countries are adopting emission protocols like cap-and-trade or alternatives to reduce greenhouse gas (GHG) emissions. They use carbon "trading" to help industries with a small "carbon footprint" and make those with higher emissions buy carbon "allowances." Concerns about the environment, fossil fuel usage, and energy conservation have led to a lot of focus on generating electricity from renewable sources. Wind and solar energy are abundant and readily available [4, 6, 7]. Interestingly, the sun's radiation energy that the earth receives in just one hour is enough to meet the world's energy needs for a whole year. Seizing a small fraction of the abundant wind energy could significantly contribute to meeting global electricity needs. While advancements in technology are still necessary to harness renewable energy efficiently, solar and wind power technologies have progressed rapidly. As of 2021, the installed wind power capacity worldwide reached approximately 1,870 GW [8]. The World Energy Council predicts that the global wind capacity could reach up to 8,000 GW by 2030 to achieve a Net Zero Emission Scenario. The deployment of photovoltaic (PV) systems is currently growing at a rate of 40% per year globally [6]. The USA aims to achieve 160 GW of solar power by 2050. However, solar and wind energy sources are not constant and reliable. The unpredictable nature of these alternative sources poses significant challenges for grid operators, as other power plants (typically fossil-fueled units) need to compensate for their variability. For instance, as revealed in Figure 1.2(a), wind power profiles in Tehachapi, California, vary over minutes, hours, and days, peaking at night when demand is low. Throughout the day, wind power can range from a few GW at certain moments to just a few megawatts (MW) or even zero at others. Similarly, in Figure 1.2(b), solar power is generated only during daylight and fluctuates when clouds pass by [9]. Another concern is that unconventional resources are limited and often located far from load centers. In the USA, wind farms are concentrated in the Midwest regions, while solar installations are prominent in southwest counties. To mitigate the variability of unconventional energy generation, affordable electrical energy storage (EES) will become essential. EES has been considered a key enabler of the smart grid or future grid, which is expected to integrate a significant amount of unconventional energy sources while providing power (i.e., electricity) to electric and hybrid vehicles [10], although the cost of implementing EES is a major concern [11].

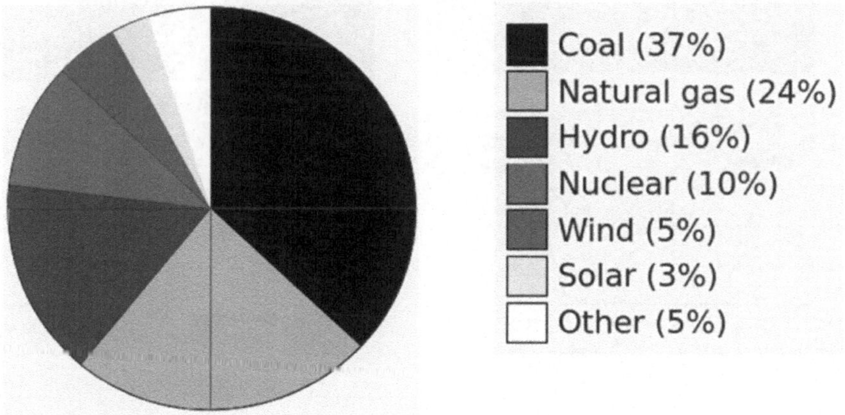

FIGURE 1.1 2019 world electricity generation by source (total generation was 27 petawatt-hours).
Source: Adapted from Ref. [2]

Nanomaterials (NMs) offer significantly enhanced ionic conductivity and electronic conduction compared to conventional battery and supercapacitor materials. They enable the utilization of all available intercalation sites within the electrode structure, resulting in high specific capacities and fast ion dispersion. These characteristics make NM-based devices capable of handling high currents, providing a promising solution for efficient energy storage and high-power applications. However, there are still several challenges associated with their implementation in energy storage technology, and apart from the use of multiwall carbon-nanotube composites and carbon coatings on silicon electrodes in lithium-ion batteries, the commercial utilization of NMs remains limited. After many years of progress, a collection of NMs with multipurpose chemical conformations and forms remains, extending from carbides, oxides, and chalcogenides to carbon and elements forming mixtures with lithium. This chapter contains several subdivision morphologies, such as zero-dimensional (0D) quantum dots and nanoparticles; one-dimensional (1D) nanotubes, nanowires, and nanobelts; two-dimensional (2D) nanosheets and nanoflakes; and three-dimensional (3D) permeable nanowebs. Alongside lithium and lithium ions, these chemically diverse nanoscale subunits are available for generating energy storage solutions, such as wearable and portable energy storage technologies, which are not achievable with traditional resources [12].

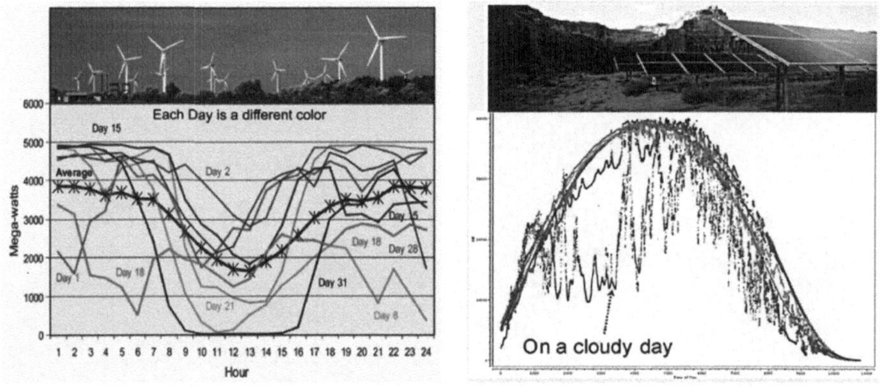

(a) (b)

FIGURE 1.2 (a) Day-to-day outlines of wind power predictable by 7 × output in April 2005 for the year 2011 in Tehachapi, California. (b) 5 MW PV power over an extent of 6 days in Spain [9].

Source: Courtesy of (a) ISO California; and (b) AES.

1.1.1 NEED FOR ENERGY STORAGE

Speculation in energy storage changes with the network, generating longstanding advantage and dependability for years to come. Energy storage is an important pivot for the complete grid, supplementing capital from renewable, conventional base-load nuclear and fossil fuels, to mandate lateral resources and scheme competence properties. It can perform as a creation, spread, or delivery asset – occasionally in a solitary asset. Eventually, storage is an empowering know-how. It can save customers' money, advance dependability and flexibility, assimilate production bases, and aid in reducing ecological impressions [13].

Energy storage can help reduce operating expenses for the power grid and save money for electricity consumers who implement energy storage systems in their homes and industries. It enables the reduction of costs associated with peak demand management and backup power generation, while also allowing the utilization of cheaper energy during off-peak periods. By having energy storage in place, businesses can avoid costly disruptions and maintain normal operations during temporary outages. Residents can safeguard their perishable food and medicines, as well as avoid the inconvenience of being without electricity. Additionally, both commercial and residential customers have the option to participate in demand response programs when available [14].

Energy storage has the potential to provide backup power during power outages. The same idea that applies to backup power for a single device (e.g., a smoulder fright that wads into a home but also possess battery backup), can be scaled up to an entire building or even the entire power grid. Storage provides the necessary flexibility to the grid, ensuring that uninterrupted power supply is available to customers whenever and wherever they need it. This flexibility is crucial for both reliability and adaptability. As the cost of power outages continues to rise, the value of improved consistency and increased resilience also increases. [15]

Energy storage can smooth out the transportation of variable or intermittent resources, such as solar and wind, by storing excess energy when the wind is blowing and the sun is shining, and releasing it when the opposite is true. However, storage can also support the efficient delivery of electricity for steady, baseload needs. When demand fluctuates rapidly and flexibility is crucial, energy storage can supply or extract electricity as needed to precisely match the load – whenever and wherever it is required. Energy storage is an enabling technology. It can be there when the sun isn't shining or the wind isn't blowing. When demand fluctuates and baseload resources can't respond quickly enough, energy storage can be beneficial [16].

In the most humble circumstances, energy storage allows for the preservation of electricity for future use, precisely when and where it is most needed. This enables the enhancement of capabilities and skills for the electric grid, including the ability to reduce greenhouse gas emissions. By introducing more flexibility into the grid, energy storage can facilitate the integration of additional wind, solar, and distributed energy resources. It can also improve the efficiency of the grid by increasing the capacity factor of existing assets and offsetting the need for constructing new, environmentally hazardous peak power stations. As the combination of energy sources becomes cleaner with low- and no-carbon resources, energy storage assists in the seamless and reliable growth of that resource mix [17].

1.1.2 ENERGY STORAGE DEVICES: BATTERY AND SUPERCAPACITOR

Battery storage is crucial for replacing hydrocarbon deposits with renewable energy. These systems play a key role in clean energy and meeting electricity demands [18]. Battery energy storage systems (BESS) allow renewable energy from sources like wind and solar to be stored and released when needed. Lithium-ion batteries, commonly used in phones and electric vehicles, are the leading technology for large-scale energy storage, ensuring a steady supply of renewable energy. This technology is essential for powering establishments with clean energy, even during periods of low sunlight or wind. For example, the UK has a significant offshore wind capacity, but effectively harnessing and managing this energy can increase its value by boosting production and potentially reducing costs. Daily, engineers at National Net and electricity networks worldwide must align supply with demand. Managing these summits and managers turns out to be more challenging when the goal is to achieve net zero carbon emissions by phasing out fossil fuel plants that have traditionally been used as a reliable, baseload source of energy [19]. The UK government estimates that technologies like battery storage systems – supporting the integration of more low-carbon power, heat, and transportation technologies – could save the UK energy system up to £40 billion (or $48 billion) by 2050, ultimately reducing household energy costs. Battery energy storage systems are significantly more advanced than small domestic batteries. A battery storage system can be charged with electricity generated from renewable energy sources, such as solar and wind power. Intelligent battery software uses algorithms to manage energy production, and automated control systems are used to determine when to store the energy for reserves or discharge it to the grid. Energy is unlimited with the battery storage system, keeping costs low and electricity stable. This text explains the extensive battery storage arrangements, but home energy storage systems work on the same principles. Storing renewable energy requires affordable technologies that can be discharged and charged millions of times, are safe, and can store enough energy economically. Lithium-ion batteries were developed by a British researcher in the 1970s and were first used commercially by Sony in 1991 for their handheld video recorder [20]. While they are now the most cost-effective energy storage solution, there are other battery storage technologies currently being developed. These include:

1. **Compressed Air Energy Storage:** With these schemes, usually positioned in large cavities, excess power is used to compress air and then store it. While energy is required, the compressed air is released and passes through an air turbine to generate electricity.
2. **Mechanical Gravity Energy Storage:** One example of this type of system occurs when energy is used to lift concrete blocks up a tower. When the energy is needed, the concrete slabs are released, generating electricity through the force of gravity.
3. **Flow Batteries:** In these batteries, which are fundamentally rechargeable fuel cells, chemical energy is supplied by two chemical constituents liquefied in fluids confined within the system and separated by a membrane.

Supercapacitors (SCs) are a type of energy storage device that surpasses both batteries and traditional capacitors. They have a higher energy storage capacity than conventional capacitors and can deliver it at a higher power output compared to batteries. These characteristics, along with their longer lifespan and excellent cycling performance, make SCs an exceptional energy storage solution [21]. They are currently used in various applications, either in combination with other energy storage strategies (often batteries) or as standalone energy sources. Porous carbons are being incorporated into the electrodes of commercial SCs due to their high conductivity and surface area. A supercapacitor's main characteristic is to accumulate energy by exciting ions in the electrolyte on the electrode shells. The original type of supercapacitor, the electrical double-layer capacitor, utilizes the reversible electrostatic buildup of ions on the surface of a porous electrode. This category includes carbon compounds with a large surface area [22]. Secondly, the pseudocapacitor category involves the reversible redox reaction with electrolyte ions, where the ions interact with surface functional groups. Pseudocapacitive energy storage consists of oxide compounds of transition metals (TMs) such as ruthenium and manganese, conducting polymers (CPs), and carbon compounds doped with different atoms [23]. The third type, a hybrid capacitor, is essentially a combination of a faradaic battery-type electrode and a non-faradaic electrical dual layer capacitor-type electrode [24].

The faradaic battery electrode is composed of phosphides, sulfides, and TM oxides, among other materials. Additionally, battery-type electrodes are often confused with pseudocapacitive electrodes, despite the fact that their charge storage mechanisms are fundamentally different. Pseudocapacitive charge storage is achieved using a highly reversible surface redox reaction that does not require phase transition. On the other hand, battery-type electrodes store charge through a reversible faradaic process where the charged and discharged electrodes undergo a phase conversion. SCs are used as standalone energy sources or in combination with other strategies. The following are some of the applications that are utilizing SCs:

1. **Transferable Strategies:** SCs are used as an energy source in portable screwdrivers and camera flashes, as they require short bursts of energy and rapid, continuous recharging.
2. **Recollection Holdups:** Service Centers (SCs) with a fast response time provide a temporary solution to a momentary interruption in the power supply of the memory.
3. **Decoupling of Energy and Power Necessities:** SCs are used to meet energy needs, whereas a dissimilar scheme delivers the main foundation of energy. Decoupling of this kind is common in hybrid and electric vehicles.
4. **Revival Strategies:** SCs can recuperate energy unconfined by machineries that accomplish monotonous and fixed activities. They are seen in a diversity of tenders, including cranes and lifts, as well as in the slowing schemes of electric or hybrid cars such as trains, buses, and trash or distribution vans.
5. **Electricity Circulation and Stowage:** SCs are proficient at recompensing for small-spell variations in voltage indications in the circulation line and for linking the break between the making and burning of electricity. Moreover, SCs are applied to resume power schemes that have been botched or to deliver energy until the actual source is restored. SCs are also utilized as energy storage devices in renewable production units, most particularly wind energy, for their low preservation necessities.

1.2 MECHANISM FOR SUPERCAPACITOR

Decrypting the charge storing machinery of conformist SCs is crucial for developing high-energy density SCs with long-lasting reusability, which can greatly mitigate energy crisis. Although ex situ characterization methods have helped understand the charge storing mechanisms of SCs, there are still many unknowns that need to be addressed. Over the past decade, in situ characterization techniques such as 'in-situ' XRD, XAS, XPS, IR/FTIR, NMR, AFM, SEM, TEM, and EQCM have emerged to shed light on the charge storage mechanisms of SCs [25]. By combining in situ methods with theoretical simulations, we can effectively explain the ion behavior and charge transfer in SC electrode systems. The following sections will discuss the most significant SC mechanisms in detail.

1.2.1 ELECTRIC DOUBLE LAYER SUPERCAPACITOR (EDLC) MECHANISM

The charge storage in EDLCs depends on the movement of ions at the electrode interface without any charge transfer reaction. EDLC-based SCs use large surface area and electronically conductive electrode materials. Figure 1.3 shows a schematic diagram of EDLCs [26]. In the 19th century, Helmholtz introduced the first prototype for EDLCs, known as the Helmholtz model (Figure 1.3(a)). This model describes the separation of opposite charges at the electrode interface. The Helmholtz model is similar to traditional dielectric capacitors. Later, Gouy and Chapman improved the Helmholtz EDLC model, known as the Gouy–Chapman theory (Figure 1.3(b)). This theory assumes that ions are point charges, and their distribution near the electrode surface follows a Boltzmann distribution. However, the Gouy–Chapman hypothesis fails to explain the highly charged double layers in the EDLC model. To address the limitations of the Gouy-Chapman model, Stern combined the Helmholtz conjecture and Gouy–Chapman hypothesis. This differentiation led to the identification of two regions of ion delivery: the internal region known as the Stern layer, and the outer diffuse layer (Figure 1.3(c)). In the Stern film, the ions are strongly attracted to the probes, which consist of a multi-layer assembly including the inner Helmholtz plane (IHP) and outer Helmholtz plane (OHP). A verbose layer is located beyond the OHP. As a result, the overall double-layer capacitance (C_{dl}) of the probe can be defined using the following equation:

$$\frac{1}{C_{dl}} = \frac{1}{C_{Stern}} + \frac{1}{C_{diff}} \tag{1}$$

where C_{Stern} represents the double-layer capacitance of the Stern layer, and C_{diff} represents the double-layer capacitance of the diffuse layer. Carbon-based

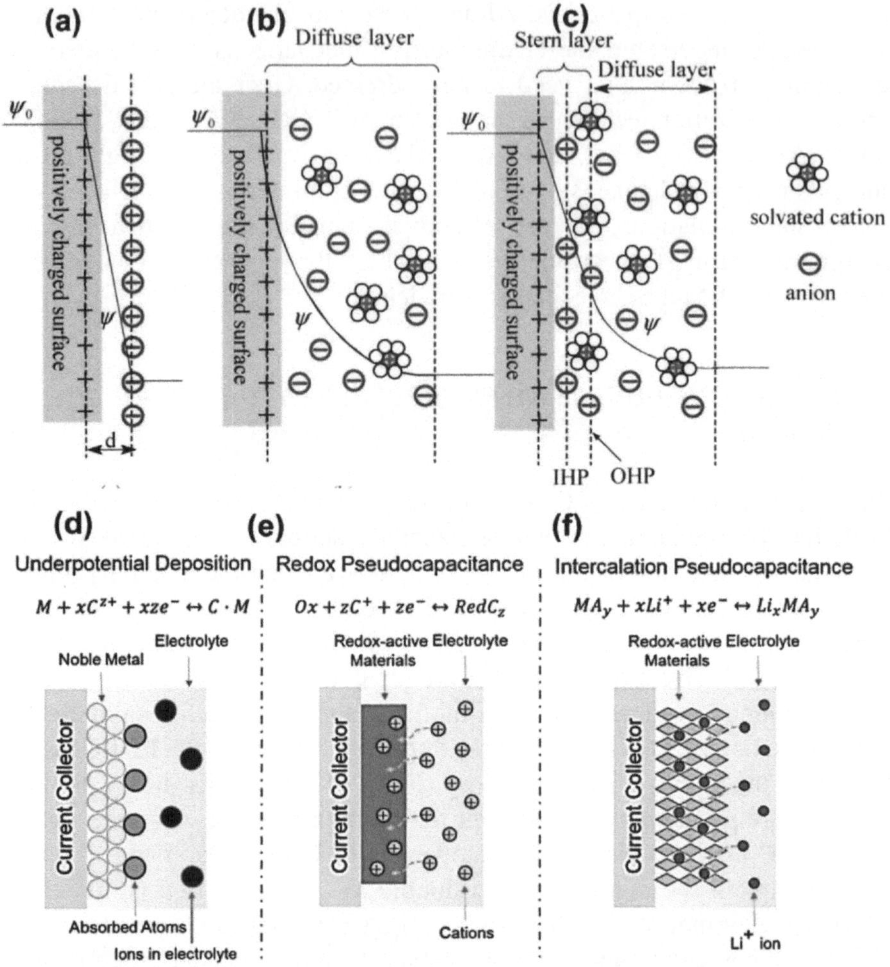

FIGURE 1.3 Models of the electrical double layer include: (a) Helmholtz model; (b) Gouy-Chapman model; and (c) Stern model. Additionally, there are various types of pseudocapacitance mechanisms, represented by schematic diagrams: (d) underpotential deposition; (e) redox pseudocapacitance; and (f) intercalation pseudocapacitance.

Source: Reproduced with permission [26].

materials such as graphene, activated carbon (AC), and carbon nanotubes (CNT) are commonly used in electric double-layer capacitors (EDLCs). AC is often chosen as a precursor material for supercapacitors due to its high surface area, good electrical conductivity, and reasonable cost. AC can be produced through the carbonization process of various carbonaceous sources such as coal, wood, and nutshell. Graphene, a single layer of graphite composed of sp^2-bonded carbon atoms arranged in a polyaromatic honeycomb lattice, is utilized in high-performance energy storage systems for its exceptional physicochemical properties, including high surface area, good electrical conductivity, excellent thermal and chemical stability, and remarkable flexibility. Characteristically, CNTs can be shaped from rolled-up graphene sheets and, depending on the fabrication process, can be categorized as SWCNTs and MWCNTs. The arrival of CNTs has unlocked new potentials in energy storage devices, leading to higher capacitance values due to their knotted webs with an exposed vital canal [27].

1.2.2 PSEUDO-CAPACITORS' CAPACITANCE

Comparing EDLCs, PCs stock charge in the probe constituents via a faradaic course, involving exterior or near-surface rapid reversible redox responses [28]. As per the reviewed theory established by Conway about the functioning machinery of PCs, the charge storage machinery of PCs can be separated as follows: (i) underpotential deposition (UD); (ii) surface redox pseudocapacitance (SRP); and (iii) intercalation pseudocapacitance (IP). The UD course comprises the faradaic adsorption or detachment of metal ions on diverse kinds of noble metals' surfaces to build an adsorbed monolayer above their redox potential (Figure 1.3(d)) [26]. SRP happens on the superficial or close to the exterior of ingredients for the associated faradaic charge-transfer course (Figure 1.3(e)) [26]. Likewise, IP designates the method of intercalation of ions into the channels or coatings of a redox-active electrode material, which is associated with the faradaic charge-transfer process without a crystallographic phase transition (Figure 1.3(f)) [26]. These three charge storage mechanisms occur in PCs due to various physical processes and with different types of electrode materials. The understanding of charge storing in PC electrodes is based on the following equation:

$$E' \sim E - \frac{RT}{nf} \ln \frac{X}{1-X} \tag{2}$$

Characteristically, metal oxides such as RuO_2, MnO_2, IrO_2, NiO, SnO_2, Co_2O_3, MoO_x, and V_2O_5, metal chalcogenides, metal hydroxides, and CP-founded probe substances are incorporated as pseudocapacitors. Typically, MnO_2 and RuO_2 are researched as pseudocapacitive probes. CPs, including polyaniline (PANI), polypyrrole, and polythiophene, are employed as pseudocapacitive probes for their high flexibility and conduction [29]. The charge-discharge profiles and cyclic voltammetry (CV) of various types of charge storage mechanisms of energy storage resources are shown in Figure 1.4(a) [26].

Type-A signifies the CV and charge–discharge profile of the capacitive probe supplies, manifesting quadrangular cyclic voltammograms and a rectilinear voltage response throughout continuous current charging/discharging. Correspondingly, type-B pseudocapacitive is faradaic in nature, while type-C faradaic is not the classical capacitive in behavior, rather it is more like a battery (Figure 1.4) [26].

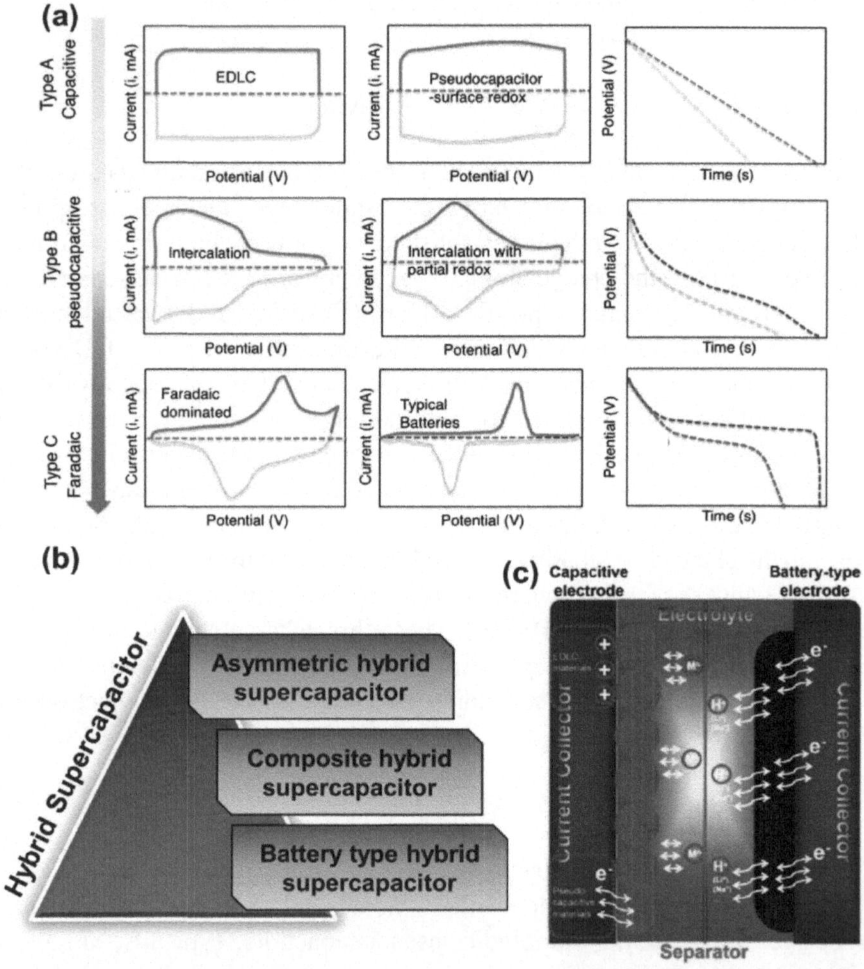

FIGURE 1.4 (a) Cyclic voltammograms and corresponding charge–discharge profiles for various types of energy storage materials; (b) classification of hybrid supercapacitors according to their design or working mechanism; and (c) schematic diagram of a battery–supercapacitor hybrid system.

Source: Reproduced with permission [26].

1.2.3 HYBRID SUPERCAPACITOR

Hybrid supercapacitors (HS) consist of polarizable electrodes (EDLC-based) and non-polarizable electrodes (PCs) to store charges. This type of SC combines the energy density of a battery-like electrode and the power density of a capacitor-like electrode in the same device, utilizing both faradaic and non-faradaic processes. Hybridization can be achieved at either the device or material level. Based on the hybridization process of SCs, HSs can be divided into three classes as follows (Figure 1.4(b) and (c)) [26]: (i) asymmetric; (ii) composite; and (iii) battery-type. Asymmetric HSs are designed using carbon-derived materials as anode and metal oxide or sulfide as cathode to achieve high energy and power density simultaneously. The properties of both carbon-based materials and chalcogenides, as well as transition metal oxides, are combined to create complex hybrid supercapacitors (HS). As a result, composite HS demonstrate better electrochemical performance, including high energy density, specific capacitance, and power density, compared to pristine EDLC-based or PCs-based supercapacitors (SCs), due to the high electric conduction, large surface area, and excellent mechanical strength of composite hybrid materials. The battery-type hybrid supercapacitor combines two different electrodes, where the battery-type electrodes undergo faradaic redox reactions during the charging and discharging processes [24].

1.2.4 SYMMETRIC AND ASYMMETRIC SUPERCAPACITOR (ASC)

Symmetric supercapacitors (SSCs) partake mutually positive and negative electrodes that are fabricated with precisely the same resources. These materials comprise metal oxides, CPs, transition metal dichalcogenides (MX_2), transition metal trichalcogenides (MX_3), MXenes, and carbonaceous substances. The charge storing machinery of these ingredients is not very complicated while compared with asymmetric supercapacitors (ASCs). The charge transference and charge stowage machinery are principally EDLC type or pseudocapacitive type, administered by the electrode substance. Sporadically, complex materials (combination of EDLC and pseudocapacitive) are utilized as the probes, the charge stowage knowhow of which is either prevailed by one of the resources or it may be a cross type. Lately, several researchers have progressed on SSC strategies to attain great capacitance along with decent cycling steadiness employing dual and treble metal sulfides and oxides and their mixtures. Isac Franklin et al. organized hydrogen-deprived carbon nanorod probes for a high energy density (22 Wh/kg @ 1 A/g) and excellent cycling stability (96.65%)-based supercapacitors. The charge storage mechanism is completely EDLC-type, exhibiting a capacity of 198 F/g in a dual-electrode configuration [31]. Mane et al. recently incorporated MnO_2/RGO electrodes to power 211 LEDs for a duration of 3 minutes, and their flexible hybrid device demonstrated a high energy and power density of 57 Wh/kg and 1.6 kW/kg, respectively. The charge storage mechanism exhibited pseudocapacitive behavior, as clearly demonstrated by the electrochemical performance [32]. An impressive energy density of 81 Wh/kg at 714 W/kg was achieved by an IrO_2@Mn_3O_4 symmetric device with rectangular-shaped CV curves, developed by Beknalkar et al. The synergistic effect of the combined oxides contributed to the high specific capacity and remarkable cycling stability [33]. Recently, numerous MXenes and metal carbides have been explored as supercapacitor materials, surpassing the majority of previously discovered substances in terms of their energy storage capability [34].

If the pair of electrodes in a supercapacitor are not identical in all aspects, such as possessing equivalent properties, mass ratio, functional clusters, charge storage mechanisms, and active redox sites, then they create an asymmetric device. Furthermore, hybrid strategies are advantageous in this case, as their two introduced electrodes (one capacitive type and another, battery faradaic type) have different charge storage schemes. The significant main advantage of the asymmetric supercapacitor (ASSC) is that it offers a wide potential operational range, leading to high energy density and power density, which is not easily achievable in the case of symmetric supercapacitor (SSC) strategies. A redox-active medium can be introduced into the symmetric system as a substitute for an inactive electrolyte to create an ASSC device if sufficient redox couples naturally occur. Recently, in the past few years, there has been an abundance of Advanced Sodium-Sulfur (ASS) plans in the pursuit of conventional energy sources. Wu et al. developed an ASSC device using cobalt nickel double hydroxide as the anode and cobalt metal–organic framework (MOF) as the cathode. This device achieved an energy density of 42.5 Wh/kg with a stability of 81.5% and a wide operating potential [35]. Ghose et al. combined exfoliated MXene and $MoS_{(3-x)}$ as the cathodes and anodes, respectively, to create a dispersion-associated pseudocapacitance apparatus with a potential range of 1.6 V, exhibiting high energy concentration and power density [36]. The comprehensive device strategy and mechanism of ASS were thoroughly evaluated by two different research groups [37, 38].

1.3 SUPERCAPACITOR BASED ON METAL OXIDE AND ITS HYBRIDS

Among the various materials used for supercapacitors (SCs), metal oxides have garnered significant attention due to their physical and chemical properties. In comparison to metal chalcogenides, metal oxides are easy to fabricate, allowing for large-scale production. They are also chemically stable, environmentally friendly, and compatible with a wide range of electrolytes, which are crucial for achieving robust and cost-effective SC strategies. Additionally, metal oxides exhibit higher theoretical specific capacitance, greater energy density, and improved electrochemical performance compared to graphene and other 2D materials. This is due to their large surface area and multiple oxidation states [39, 40]. A wide range of metal oxides, including single metal oxides such as TiO_2, RuO_2, Co_3O_4, MnO_2, V_2O_5, NiO, Cr_2O_3, and MoO_3, as well as dual metal oxides like $NiCo_2O_4$, CuCo$_2$O4, $FeCo_2O_4$, $ZnCo_2O_4$, etc., have been effectively prepared through direct exfoliation from their corresponding bulk counterparts or self-assembly after salt precursor-based mixtures [41]. For example, manganese oxide (MnO_2) and ruthenium oxide (RuO_2) have been extensively studied as electrode materials for supercapacitors due to their fast, reversible redox reactions [42, 43].

Typically, the supercapacitive performance of metal oxides depends on several critical aspects, such as (i) crystallinity; (ii) morphology; (iii) specific surface area; (iv) electronic conduction; and (v) mass concentration of the electroactive substance [44, 45]. In general, amorphous metal oxides demonstrate superior electrochemical properties compared to their highly crystalline counterparts [46, 47]. Regarding 2D assemblies, such as nano-lamellar, nanosheet, and nano-film structures, SCs exhibit improved physical and chemical attributes, including mechanical stability, flexibility, transparency, and chemical stability, when compared to other nanostructures (0D and 1D) [48]. An electrode with a high specific surface area can enhance electrochemical performance in terms of both specific capacity and energy density, allowing for more electroactive sites and facilitating multiple redox reactions [49]. Most metal oxide complexes, including 0D, 1D, and 2D materials, exhibit poor electronic conduction, which hinders electron and ion transport during the electrochemical charge–discharge process. At high charge–discharge rates, the impedance to electronic movement becomes significantly high for electroactive substances like metal oxides, resulting in reduced specific capacity, energy density, power density, and rate capability [46]. A high mass loading of electroactive materials (metal oxides) leads to decreased electrical conduction and restricted access to electrolyte ions, resulting in poor electrochemical performance. In essence, the mass loading factor is inversely proportional to the specific capacity, meaning that an increase in mass loading would decrease the specific electrical capacity [50, 51].

Even though metal oxides possess fascinating electrochemical performances based on the faradic redox response, they may not be utilized in their pristine form as supercapacitor (SC) electrodes due to the following limitations: (i) the majority of metal oxides exhibit poor electrical conduction, resulting in increased resistance and charge transport hindrance, leading to significant voltage drop at high current densities during charge–discharge cycles. This negatively impacts the rate capability and power density of SC devices; (ii) during cycling of metal oxide-based SCs, stress accumulates on the surfaces of metal oxides, causing structural degradation of the electrode material and resulting in poor cycle life; and (iii) it is challenging to modify certain properties of metal oxides, such as increasing the specific surface area (SSA) and achieving uniform pore size distribution. These are the main disadvantages of metal oxide-based SCs [52]. To overcome these shortcomings and enhance their electrochemical performance, it is essential to incorporate emerging complexes and/or cross-probe constituents by combining metal oxides with carbonaceous materials or conductive polymers [53].

1.3.1 METAL OXIDE-GRAPHENE CROSSES

Usually, graphene amalgams of metal-oxides have numerous advantages in supercapacitor applications for the following reasons: (i) the metal oxide/graphene composite structure can greatly enhance electron conduction due to the graphene coating on the metal oxides, increasing the hole density in graphene [54]; (ii) metal oxide-graphene composites have a large surface area, which improves the interaction between the electroactive material and the electrolyte, resulting in enhanced electrochemical performance [55]; and (iii) in metal-oxide graphene hybrids, metal-oxide nanosheets are often sandwiched between graphene nanosheets to prevent the aggregation of graphene layers. Zhang et al. fabricated a 2D MnO_2/graphene composite using the electrostatic coprecipitation process [56]. The morphological investigation (Figure 1.5(a) and (b)) of the 2D MnO_2/graphene mixture reveals that the MnO_2 nanosheets are distributed on the surface of the graphene substrate due to electrostatic interaction. This 2D MnO_2/graphene hybrid exhibits enhanced electrochemical performance compared to pristine

MnO_2 and graphene, thanks to the synergistic effect of the excellent electrical conductivity of graphene and the pseudocapacitive properties of the MnO_2 nanosheets. Figure 1.5(c) illustrates the CV profile of the 2D MnO_2/graphene composite, achieving a specific capacity of approximately 188 F/g at a current density of 0.25 A/g. In a similar manner, Ge et al. organized a NiO-graphene mixture through a simple hydrothermal process [57]. SEM images of plain NiO nanocoats and the NiO-graphene amalgam are shown in Figure 1.5(d) and (e). The electrochemical characterization of graphene, NiO nanocoat, and the NiO-graphene composite probes was conducted in a three-electrode setup in a 2 M KOH aqueous medium. The NiO-graphene composite probe achieved a specific capacity of approximately 348 F/g, which is higher than the values obtained for both pure NiO and graphene alone. Figure 1.5(f) demonstrates the cycling stability performance of the NiO-graphene composite probe over 500 cycles at a current density of 1.5 A/g. To fabricate a flexible planar supercapacitor, Peng et al. utilized the hybrid nanostructures of quasi-2D ultra slim MnO_2 graphene nanosheets. The two-dimensional supercapacitor device exhibited a specific capacity of 267 F/g at a current density of 0.2 A/g with exceptional cycling stability (92% capacitance retention even after 7,000 charge–discharge cycles) [58]. Besides, Tan et al. equipped a 2D Co_3O_4 nanoflake-graphene mixture on a Ni foam substrate by CVD method followed by hydrothermal treatment and achieved a high capacity of 1.75 F/cm^2 at a current density of 1 mA/cm^2 [59]. Recently, Shrestha et al. synthesized $MnCo_2O_4$/nitrogen-doped graphene 2D/2D hybrid nanostructures using a one-pot polyol process and performed electrochemical analysis by fabricating an asymmetric device [60]. Figure 1.5(g) illustrates the SEM image of the $MnCo_2O_4$/nitrogen-doped graphene 2D/2D cross structure, which revealed that the interaction of the 2D $MnCo_2O_4$ nanoflakes with the planar nitrogen-doped graphene (with a crumpled structure) prevents the reorganization of nitrogen-doped graphene. Electrochemical performance of the bare $MnCo_2O_4$ and $MnCo_2O_4$/nitrogen-doped graphene 2D/2D cross probe was evaluated in a three-electrode system with 2 M KOH as the aqueous medium, and the corresponding CV curve at a scan rate of 10 mV/s is shown in Figure 1.5(h). The CV contour reveals that the $MnCo_2O_4$/nitrogen-defected graphene cross probe exhibits improved electrochemical activity compared to the pristine $MnCo_2O_4$ probe, thanks to the synergistic interaction between the spinel planar $MnCo_2O_4$ and the layered nitrogen-doped graphene. The $MnCo_2O_4$/nitrogen-decorated graphene cross-facilitated asymmetric maneuver achieves a high energy density of approximately 48.5 Wh/kg and a power density of around 16,000

W/kg. Figure 1.5(i) illustrates the cycling stability performance of the asymmetric device over 10,000 cycles, with the device retaining at least 85.9% of its initial specific capacity after 10,000 charging-discharging cycles.

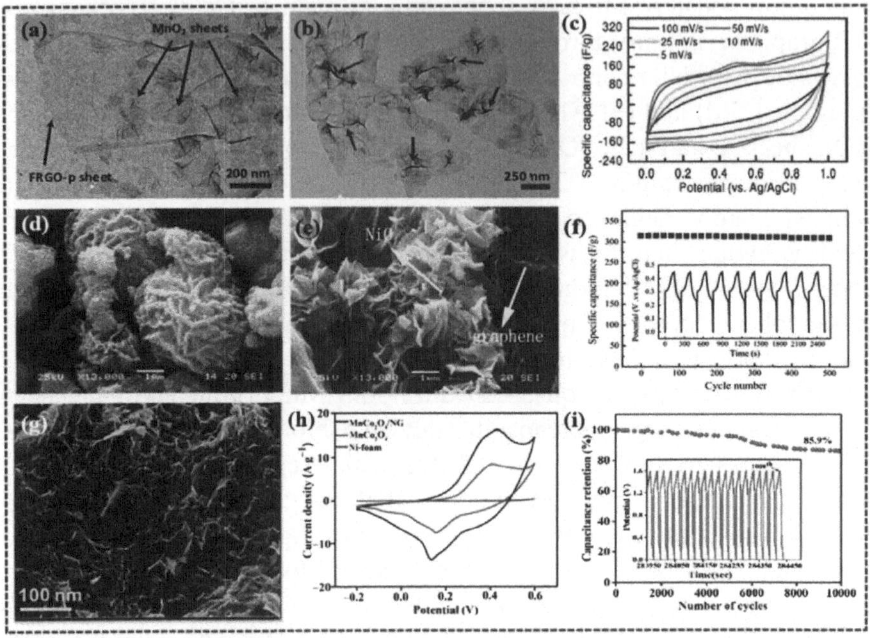

FIGURE 1.5 (a) and (b) TEM images of the 2D MnO_2/graphene cross at diverse amplifications; and (c) CV outlines of the MnO_2/graphene mixture at several scan paces. SEM photograph of (d) plain NiO nanosheets; (e) the NiO/graphene mix; and (f) the cycling steadiness silhouette of the NiO/graphene mixture probe. (g) SEM photo of the $MnCo_2O_4$/nitrogen-doped graphene 2D/2D amalgam nanostructure; (h) the CV contour of plain nickel foam, MnO_2O_4, and $MnCo_2O_4$/nitrogen-doped graphene 2D/2D cross electrodes; and (i) the cycling constancy curve of the $MnCo_2O_4$/nitrogen-doped graphene 2D/2D hybrid-based asymmetric expedient.

Source: Reproduced with permission [53].

1.3.2 METAL OXIDE-CNT CROSSES

Carbon nanotubes (CNTs) are extensively discussed as substrates or development templates for supporting metal-oxides in supercapacitor (SC) and lithium-ion battery (LIB) applications due to their high electrical conduction, thermal stability, and structural firmness [61]. Zhou et al. prepared various transition-metal-oxide/CNT composites using a wet chemical technique followed by annealing (as shown in Figure 1.6(a)) for both SC and LIB applications [62]. The SEM image in Figure 1.6(b) illustrates the nanosheet structures of $NiCoO_2$ on the CNT backbone. The electrochemical characterization of the $NiCo_2O_4$/CNT composite was performed in a three-electrode setup with a 2 M KOH electrolyte. The $NiCo_2O_4$/CNT composite electrode achieved a high specific capacity of approximately 1,360 F/g at a current density of 10 A/g with excellent cycling stability (3,000 cycles). Similarly, Mondal et al. synthesized hybrid MWCNT/$NiCo_2O_4$ nanosheets using a microwave-assisted technique followed by calcination in air [63]. Figure 1.6(c) and (d) demonstrate the SEM and TEM photos of the MWCNT/$NiCo_2O_4$ cross, indicating that the $NiCo_2O_4$ nanosheets have grown around the exterior of the MWCNTs in the mixed MWCNT/$NiCo_2O_4$ nanosheets. For SC investigation, the electrochemical characterization was conducted in a 2 M KOH aqueous medium. The CV profile of the MWCNT/$NiCo_2O_4$ nanosheet electrode is shown in Figure 1.6(e), and the MWCNT/$NiCo_2O_4$ cross electrode exhibited a superior specific capacity of approximately 1,395 F/g at a nominal current of 1 A/g. The bi-electrode electrochemical analysis for MWCNT/$NiCo_2O_4$ nanosheets was performed in a coin cell configuration.

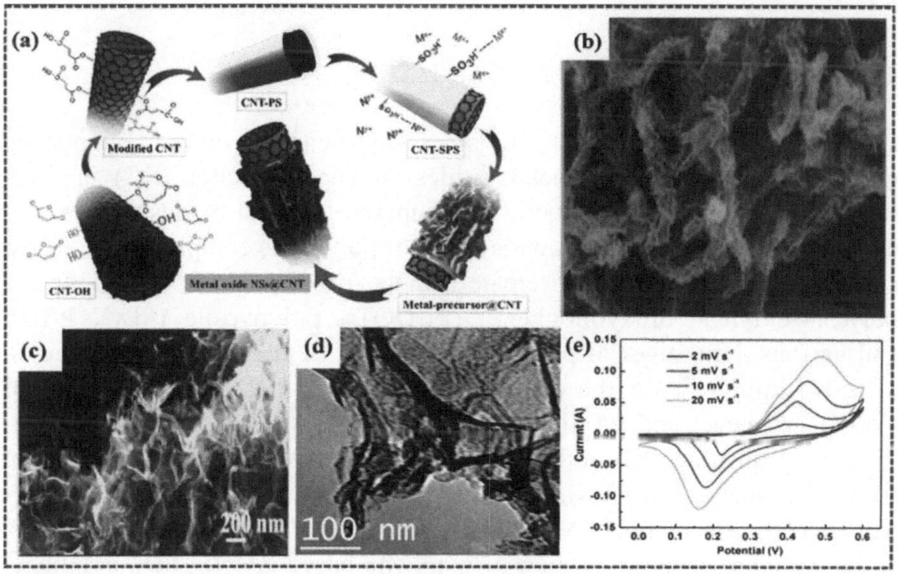

FIGURE 1.6 Representative illustration for (a) the preparation approach adapted for the synthesis of the transition metal oxide/CNT hybrid; and (b) the SEM image of the $NiCoO_2$/CNT cross. (c) and (d) SEM and TEM photos of the MWCNT/$NiCo_2O_4$ mix, respectively; and (e) CV outline of the MWCNT/$NiCo_2O_4$ hybrid probe.

Source: Reproduced with permission [53].

1.3.3 METAL OXIDE-ACTIVATED CARBON (AC) CROSSES

Activated carbon (AC) is a widely used carbon-based material for producing metal-oxide/AC composites for supercapacitor (SC) applications. AC has a low electronic conductivity and a large surface area, which provides a high actual surface area for the growth and support of metal-oxide in metal-oxide/AC composites [64]. Zhang et al. developed a 2D-coated carbon-metal-oxide composite electrode for SC applications [65]. Figure 1.7(a) shows the schematic illustration for the fabrication of the 2D-thin carbon–metal-oxide composites using a three-step process. First, a cobalt coating is electrolytically deposited on nickel foam, then, further electro-deposition is made to accumulate embedded MOFs. The SEM image of the thin carbon-metal oxide composite is shown in Figure 1.7(b), demonstrating the formation of the composite's 2D-coated structure and hexagonal shape. The electrochemical performance of the thin carbon-metal oxide composite was evaluated in 4 M KOH at scan rates between 10 and 50 mV/s, as shown in Figure 1.7(c). The coated carbon–metal oxide composite achieved a maximum areal capacitance of nearly 2,098 mF/cm^2 at a current density of 1 mA/cm^2.

1.3.4 METAL OXIDE-POLYMER CROSSES

Conducting polymers (CPs) have gained significant attention as supporting materials for embedding metal oxides in supercapacitor (SC) applications due to their high conductivity compared to metal oxides and ease of construction procedures for a wide range of polymeric complexes. The most commonly used polymers to create a metal oxide/polymer composite are poly(3,4-ethylene dioxythiophene) (PEDOT), polypyrrole (PPy), PANI, and various derivatives of polythiophene [66]. Li et al. have developed an effective approach for the controlled synthesis of 2D ternary hybrids of Co_3O_4, graphene, and PANI for SC applications [67]. Figure 1.7(d) schematically illustrates the entire formation process of 2D ternary blends of Co_3O_4, graphene, and PANI. Initially, graphene nanocoats were chemically activated by associating PANI through an on-site polymerization process, which resulted in the formation of graphene-based PANI. Then, graphene-PANI nanocoats were combined with the corresponding metal salts as precursors of metal oxides/hydroxides, followed by hydrothermal treatment to produce a ternary composite. Figure 1.7(e) shows the CV profile of the ternary hybrid electrode in the voltage range of 0.0–0.4 V at various scan rates. The synthesized ternary composite achieved a maximum capacitance of approximately 1,000 F/g at a scan rate of 5 mV/s.

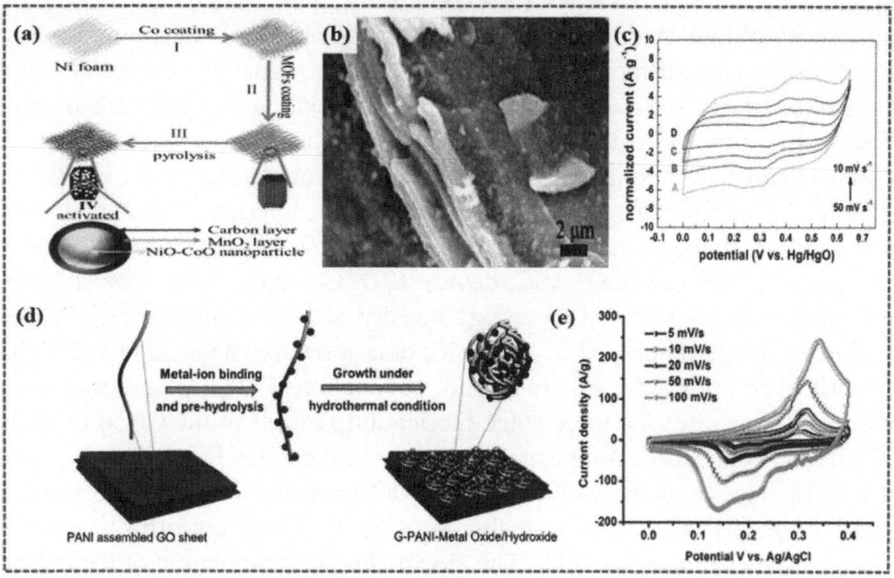

FIGURE 1.7 (a) Picture relating to the creation of the 2D-layered carbon-metal oxide cross; (b) SEM photo of the 2D-layered carbon-metal oxide mix noted at numerous enlargements; and (c) CV outline of the 2D-layered carbon-metal oxide cross at several probe rates. (d) The schematic representation of the synthesis of the 2D ternary cross of Co_3O_4, PANI, and graphene. And (e) the CV outline of the ternary hybrid quantified in a 6 M KOH watery medium.

Source: Reproduced with permission [53].

1.3.5 METAL OXIDE-METAL OXIDE CROSSES

The metal oxide/metal oxide cross-probe has a multifaceted construction that allows for rapid electron and ion transmission pathways and numerous electroactive superficial places for electrochemical response. This results in a greater specific capacity and energy compactness than the sole constituent electrode. Co_3O_4 nanowire@2DMnO_2 super-thin nanosheet mixture core/shell arrangements were developed by Liu et al. for SC usage through hydrothermal techniques. The creation of the electrode is illustrated in Figure 1.8 (a). The nanostructure was shaped with a bi-step hydrothermal process followed by an annealing course. The SEM photographs in Figure 1.8 (b) depict that the ultrathin sheets are interrelated to each other, creating an extremely permeable surface structure. The cross-arrangement displays a peak specific capacity of around 480 F/g at a flow concentration of 2.67 A/g with outstanding capacitance retention (2.7% holding even upon 5,000 sequences). Kuang et al. confirmed the manufacture (as exposed in Figure 1.8(c)) of 2D $CuCo_2O_4$ nanosheet@MnO_2 nanoflake core/shell arrangements on Ni foam and used that for ASC compartments. The hybrid arrangements are grown on a Ni foam substrate using mixed nucleation procedures associating heat and water. Figure 1.8(d) shows the SEM images of 2D $CuCo_2O_4$ nanosheet@MnO_2 nanoflake core/shell structures synthesized at various reaction periods of 24 hours. An ASC cell was fabricated using $CuCo_2O_4$@MnO_2 hybrids as the anode and activated graphene as the cathode, along with the separator and 1 M Na_2SO_4 electrolyte. The resulting ASC device (2D $CuCo_2O_4$ nanosheets@MnO_2 nanoflakes//AG) exhibited an energy density of approximately 43.3 Wh/kg within a voltage range of 2 V. Anon, Xu et al. prepared a $CuCo_2O_4$@2DNiO nanosheet composite for supercapacitor applications [70]. Figure 1.8(e) schematically illustrates the three-stage fabrication process of the $CuCo_2O_4$@2D metal oxide nanosheet composite on a Ni foam substrate. Figure 1.8(f) shows the SEM images of the $CuCo_2O_4$@2D NiO nanocoat composite, revealing that NiO sheets (with an average thickness of 50-80 nm) uniformly cover the entire surfaces of the $CuCo_2O_4$. The electrochemical evaluation of the hybrid electrode was conducted in a three-electrode configuration using 1 M KOH electrolyte, which demonstrated a high areal capacity of approximately 2.3 F cm^{-2} at a current density of 2 mA cm^{-2}, along with excellent cycling stability. Additionally, an ASC was fabricated using the $CuCo_2O_4$@2D NiO nanolayer as the anode and activated carbon as the cathode, achieving a remarkable energy density of 38.9 Wh/kg within a voltage range of 1.5 V.

1.3.6 SYNERGIC PARAPHERNALIA IN THE MIX EDIFICE

The harmonious impacts in cross assemblies (metal-oxide joint with graphene, or CNT, or AC or CP) can be momentarily emphasized as follows: (i) graphene may act a dynamic role in metal oxide/graphene cross constructions, as graphene delivers a 2D conducting pattern to indorse an interrelated conducting permeable web that would improve the electrical conduction and charge conveyance trails of metal-oxides; (ii) graphene founds a decent face to face communication and electrical connection amid itself and metal-oxides for the obtainability of oxygen-bearing efficient clusters; (iii) metal-oxides while joint with AC, and/or CNT, incline to improve the specific superficial area of cross electroactive substance, which will ease the shallow responses, and the medium can partake a decent collaboration with the probe (for its huge surface area), leading to heightened electro-chemical presentations in terms of specific capacity, energy compactness and power concentration; and (iv) for advancing the electrical conduction of plain metal-oxides, they are typically joint with numerous different CPs to augment the electrical conduction of the metal-oxide/polymer crossbred and off course to advance the numerous attributes, e.g., flexibility, specific capacity, and comfort of manufacture for real-world applications (Table 1.1).

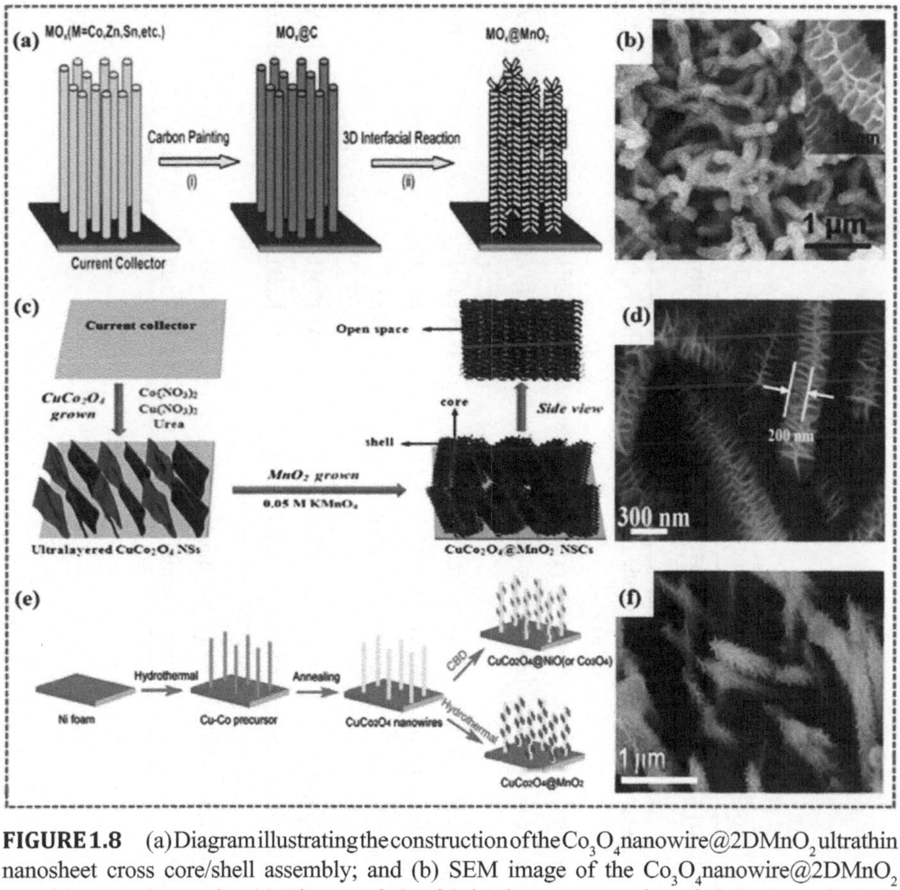

FIGURE 1.8 (a) Diagram illustrating the construction of the Co_3O_4 nanowire@2DMnO_2 ultrathin nanosheet cross core/shell assembly; and (b) SEM image of the Co_3O_4 nanowire@2DMnO_2 ultrathin nanosheet mix. (c) Picture of the fabrication process of graded $CuCo_2O_4$@MnO_2 nanostructures; and (d) SEM image of 2D $CuCo_2O_4$ nanosheet@MnO_2 nanoflake core/shell arrangements set under diverse investigational circumstances. (e) Pictorial representation for the synthesis of the $CuCo_2O_4$@2D metal oxide nanosheet cross; and (f) SEM image of the $CuCo_2O_4$@2DNiO nanosheet amalgam.

Source: Reproduced with permission – (a) and (b) [68]; (c) and (d) [69]; (e) and (f) [70].

TABLE 1.1 Summary of Energy Storage Performance of Metal Oxides with Different Hybrid Structures [53]

SL. No.	Electrode Material	Synthesis Method	Electrolyte	Specific Capacitance	Current Density	Potential Window (V)	References
1.	2D MnO_2/graphene hybrid.	Electrostatic coprecipitation method.	1 M Na_2SO_4	188 F/g	0.25 A/g	1 V	[56]
2.	2D NiO/graphene hybrid.	Chemical precipitation method.	2 M KOH	346 F/g	1.5 A/g	0.5 V	[57]
3.	2D MnO_2/graphene hybrid.	Solution treatment method.	PVA/H_3PO_4 gel electrolyte.	267 F/g	0.2 A/g	0.7 V	[58]
4.	2D Co_3O_4/graphene hybrid.	Hydrothermal method.	6 M KOH	1.75 F/cm^2	1 mA/cm^2	0.5 V	[59]
5.	2D $MnCo_2O_4$/graphene hybrid.	Polyol method	2 M KOH	1,170 F/g	1 A/g	0.5 V	[60]
6.	2D MnO_2/graphene hybrid.	Electrodeposition method.	1 M KCl	328 F/g	0.5 mA/cm^2	1 V	[71]
7.	2D $NiCoO_2$/CNT hybrid.	Wet chemical method.	2 M KOH	1,360 F/g	10 A/g	0.5 V	[62]
8.	2D $NiCo_2O_4$/MWCNT hybrid.	Microwave followed by calcination method.	2 M KOH	1,395 F/g	1 A/g	0.4 V	[63]
9.	2D Co_3O_4/PANI/graphene hybrid.	Hydrothermal method.	6 M KOH	1,063 F/g	1 A/g	0.4 V	[67]
10.	2D $CuCo_2O_4$/MnO_2 hybrid.	Hydrothermal method.	1 M Na_2SO_4	416 F/g	1 A/g	1 V	[68]

1.4 SUPERCAPACITOR BASED ON METAL DICHALCOGENIDES AND ITS HYBRID

Apart from the meaningly enhanced end proportion and superficial acquaintance from their wholesale stages, 2D TMDs partake in graphene-alike reedy constructions (~1 nm profuse for a single layer) with perfect super brief electron conveyance distance and trivial ion transportation length for rapid electrochemical courses. Research has recommended that 1H/2HTMDs, e.g., shrill MoS_2 layers, have an analogous supercapacitor presentation to that of CNT arrangements, with a theoretic capacity (~1,000 F/g) better than graphite [72, 73]. Yet, the progress of 2D TMD-founded SCs has not been efficacious since the little electrical conduction of maximum steady semiconducting 1H/2HTMDs. By distinction, metallic 1TTMDs are hypothetically metastable and partake at the basal level with further chemical bustle over that of 1H/2H stages [74–76]. They possess very tall electrical conductance (e.g., 10–100 S/cm of single-layered 1TMoS_2 layer at ambiance, 107 times more conducting than 1HMoS_2 single layers), which is on par with the excellent RGO sheets (~100 S/cm) [77]. Fresh advancements have suggested that structures of certain metallic 1TTMDs are highly stable in specific electrochemical processes [78–80], which highlights the tremendous potential of 2D TMDs in SCs. This promising prospect has been demonstrated by Chhowalla et al., where single-layered 1TMoS_2 sheets (~70% segment proportion, Figure 1.9(a)) were exfoliated from their bulk flakes using organolithium [77]. Thanks to their excellent electrical conduction, thin flexible SCs can be manufactured effectively using these 1TMoS_2 panels (stacking depth up to 5 μm, Figure 1.9(b)) without the need for any additional mediators or conductive additives (e.g., the film on polyimide substrate in Figure 1.9(d)) [77]. The resulting supercapacitor from this flexible approach exhibited volumetric capacities ranging from 400 to 700 F/cm^3 in various aqueous media (H^+, K^+, Na^+, Li^+), along with very high, stable (95% capacitance retention after 5,000 cycles) energy and power densities in organic media (the shaded and tinted area in Figure 1.9(e)). The CV arcs in the castoff voltage span (watery medium and three probes: -1.0-0 V vs. open circuit potential (OCP); biological electrolyte and two rods: 3 V) were mostly capacitive, as highlighted [77], resulting in significant power and energy concentrations (Figure 1.9(e)). The energy storage capacity in organic medium is significantly larger than that of MXenes (2D transition metal carbides, carbonitrides, and nitrides), graphene, AC, and a few high-performance Li-ion batteries (Figure 1.9(e)). The effective storage of organic electrolyte ions (e.g., the tetraethyl ammonium/TEA^+) and larger ions beyond Li^+ (e.g., Na^+, K^+) is mainly due to the disrupted sheets in loaded leaves with interlayer gaps of about 6.15 and 11.65-12.15 Å, which closely match the dimensions of these ions (Figure 1.9(c)). Following this observation, other 2D TMDs with high electrical conduction (e.g., $1TWS_2$ and $1TVS_2$ sheets) and similar large interlayer gaps are also expected to be highly effective in the construction of high-performing SCs.

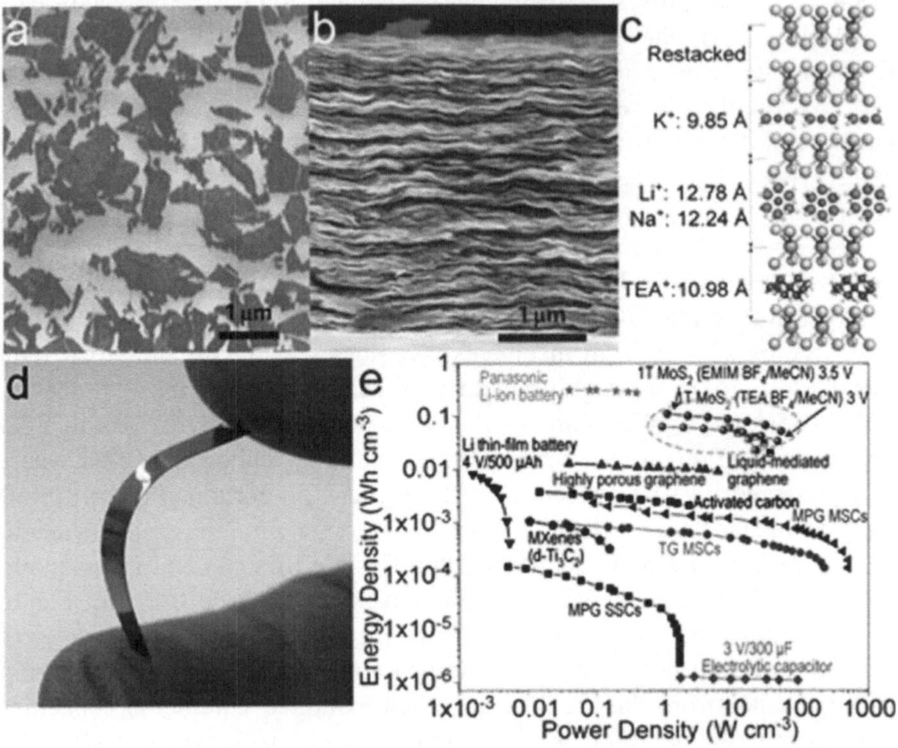

FIGURE 1.9 (a) and (b) SEM images of the disseminated single layered nanosheet 1T phase MoS_2: (a) the cross-section of the loaded $1TMoS_2$ probe. (c) Diagram displays the interlayer gap of restacked $1TMoS_2$ nanosheets (kept as 6.15 Å) earlier and later ions intercalation operation. The K^+ and Na^+/Li^+ ions are harmonized with two and six H_2O molecules, correspondingly. The intercalation of organic electrolyte (TEA+) provided an extra layer parting of 4.83 Å. (d) Image of the $1TMoS_2$ sheets on lithe polyimide substratum. And (e) Ragone plot of the bulk power and energy concentrations of diverse ingredients.

Source: Reproduced with permission [77].

Apart from the aforementioned 1TTMDs, recently developed triplex sulfides, such as Ni-Co-S [81–84], have shown great potential in SCs due to their excellent stability, favorable redox properties (compared to the respective dual Ni or Co sulfides [81]), and outstanding electrical conductivity. For instance, $NiCo_2S_4$ exhibits an electrical conductivity of approximately 10^3–10^4 S/cm at room temperature [82, 83], which is 1–2 orders of magnitude higher than that of $1TMoS_2$ single layers (10–100 S/cm) [77]. SCs incorporating these nanostructured triplex sulfides (e.g., nanofibers, 3D structures) have been extensively studied by various research groups [85–87]. While the fabrication of ultrathin single-layer ternary sulfides remains challenging, recent research on $CoNi_2S_4$ nanosheets (thickness: about 20 nm; 2D size: micrometers) has demonstrated a high capacitance of 1,117 F/g (at 2 A/g, using a three-electrode setup, 6 M KOH) and decent rate performance (1,002 F/g at 10 A/g) [88]. An ASC created after these triple sulfides offered great power (e.g., 8 kW/kg at an energy compactness of 55.4 Wh/kg) and outstanding cycling performance (>80% capacity retained after 10,000 cycles at 10 A/g) [88]. A higher capacity (2,906 F/g) than the established value of comparable $CoNi_2S_4$ layers (depth: about 45 nm; 2D dimensions: micrometers) was similarly achieved in a supercapacitor at a current density of 5 mA/cm^2 [89]. These achievements are significantly greater than those of their numerous counterparts with conventional configurations [88, 89].

Like the slim 2D TMD ingredients with deprived electrical conduction (mainly the semiconducting 1H/2HTMDs), cross assemblies after these TMDs and graphene (Figure 1.10 [90], e.g., via physical piling or chemical attachment) might be well-thought-out to hypothetically circumvent numerous faintness of old-style HS, offering better energy and power concentrations, extended cycle lifetime and enhanced protection. Contrasting conventional nanoparticles (where only the topmost exterior takes part in the electrochemical course), the entire 2D TMDs (e.g., single layers) and their extremely active ends in this cross assembly (Figure 1.10) might contribute to both the Faradaic and Non-Faradaic capacity, improving ingredients' usage and reducing cost. Contrasting it with old-style secondary batteries, the close surface redox response is fundamentally dissimilar and related to much greater energy transfer competences. Additionally, 2D TMDs might closely communicate with the graphene wall, which would safeguard fast charge transference amongst the probe and the chemically vigorous site (e.g., lively places on 1H semiconducting TMDs) for big power compactness and better rate presentation. This construction is further capable as soon as extra electrically conductive TMDs nanocrystals (e.g., 1T nanocrystals), chemical alterations, and 3D-G with huge surface parts are presented. Even though cross electrodes since sternly single-layered 2D TMDs and graphene yet require improvement, current information on alike mix constructions have established that there is a prosperous forthcoming for 2D TMDs in SCs [91–93]. For instance, Ajayan et al. have established the mix probe of $CoNi_2S_4$-graphene-layered $MoSe_2$ with graphene and slim $MoSe_2$ leaves are mutually devoted, and extremely conducting $CoNi_2S_4$ nanoparticles on/among the coatings [93]. This probe showed huge capacity (1,141 F/g, 1 A/g, tierce electrodes system, in 6M KOH electrolyte), decent rate (961 F/g at 5 A/g), and cycling presentations (108% capacity holding after 2,000 sequences, 20 A/g). The supercapacitor accumulated since this probe also displayed great power concentrations (e.g., ~0.89 and 13.3 kW/kg at energy masses of 38.6 and 23.7 Wh/kg). The part of 2D TMDs in SCs is more reinforced, predominantly when the constituents manufacture on numerous 2D nanostructures has been demonstrated to be progressively fruitful [94–102].

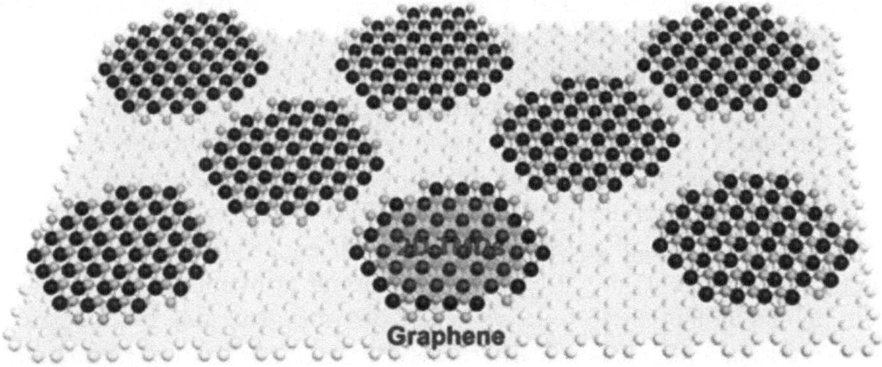

FIGURE 1.10 The diagram demonstrates the cross-assembly of 2D TMDs and graphene, where the 2D TMDs closely connect the graphene's surface with their basal planes.

Source: Reproduced with permission [90].

1.5 SUPERCAPACITOR BASED ON MXENE AND CROSSES

Since the electrode's substance plays a critical role in defining its electrochemical presentation, better materials with exclusive attributes are indispensable for the manufacture of competent probes. The unearthing of hydrous ruthenium-oxide [103] presented excellent electrochemical characteristics, including a capacity of 1,000–1,800 F/cm³ for profuse probes (>20 μm) [104] and a capacitance of nearly 2,800 F/cm³ for thin flicks (50–100 nm) [105]. Nevertheless, its great price and diminished charge stowage competence with the growing probe depth discarded it as an appropriate solid for energy stowing needs [106]. After RuO_2, the encounter of 2D TMDs [107], MOFs [108], graphene [109], etc., presented energy stowage ability by rescindable superficial redox responses. The incrusted construction of 2D substances offers a larger surface-to-volume ratio, thus guaranteeing a larger contact of electrolytes to the face of the electrode. Yet, these substances exhibited restrictions for their poorer conduction, feeble van der Waals attachment, and hydrophobicity as well. To enhance capacity, various layered resources were examined to find a suitable substitute. This led to the discovery of MXene, a 2D material with interlayer construction, a high face-to-bulk ratio, and abundant electrochemically active sites for energy storage [110, 111]. The wide range of oxidation states offered by transition metal atoms enables MXene-based supercapacitors to exhibit excellent faradaic charge storage capabilities. Unlike other pseudocapacitive materials, MXenes can store and transport charges at a significantly faster rate. Additionally, the exceptional conductivity and hydrophilic nature of MXenes make them attractive for electrochemical energy storage applications [112]. However, the use of pristine MXene as an electrode material is limited due to its tendency to restack and its inferior mechanical properties. Therefore, combining MXene with different electrode materials can greatly enhance the electrochemical performance [113]. With its excellent elastic moduli, adjustable surface functional groups, and high length-to-width ratio, MXene can be utilized in various composites. These unique characteristics of MXene composites have prompted scientists to focus on developing flexible, wearable, and portable electrochemical devices.

1.5.1 MXENE-GRAPHENE FOUNDED SUPERCAPACITORS (SCS)

The MXene-founded fibers created by the wet-spinning assemblage technique by MXene panes and RGO liquid crystal displayed an electrical conduction of 2.9×10^4 S/cm with a better areal and bulk capacity of 372.2 mF/cm^2 and 586.4 F/cm^3, respectively, in PVA/H_3PO_4 gel medium with a capacity holding beyond 3,000 sequences [114]. Kumar et al. produced a Ti_3C_2Tx MXene/graphene/Ni supercapacitor probe with the premium supernatant of HCl/LiF-imprinted Ti_3C_2Tx MXene emulsion as the substance of the probe and CVD-developed graphene on Ni-foil as the current accumulator. The manufactured supercapacitor showed a superior gravimetric capacity of ~542 F/g at a probe rate of 5 mV/s, i.e., 1.5 times bigger than the capacity attained by shorn graphene [115].

1.5.2 MXENE-CNT FOUNDED SUPERCAPACITORS (SCS)

Zhao et al. [116] described that the sandwich-like MXene/CNT complex, shaped by discontinuous percolation of MXene and CNT scatterings, presented an enhanced bulk capacity of 350 F/cm^3 at 5 A/g, in contrast with pristine MXene or haphazardly assorted MXene/CNT papers. A greater bulk capacity of 435 F/cm^3 was detected for MXene/rGO complex, with outstanding cycling steadiness. Park et al. [117] utilized 100 μm profuse MXene/CNT positive and MnO_2/CNT negative probes, joint with PVA/LiCl gel medium, to arrange an asymmetric yarn supercapacitor that presented a concentration of energy of 100 $\mu Wh/cm^2$. The Ti_3C_2Tx MXene-CNT compound probe, made using the photolithography method and purified through blankness, exhibited a superior areal capacity of 61.38 mF/cm^2 at 0.5 mA/cm^2 for CNT intercalation between MXene sheets, providing a favorable pathway for ion dispersal [118]. The Ti_3C_2Tx MXene-multiwalled CNT compound negative probe and polypyrrole-covered MCNT positive probe, combined with Na_2SO_4 medium, were used to create an ASC. This ASC demonstrated an aerial capacity of 0.94 F/cm^2 within a potential range of 0-1.6 V [119].

1.5.3 MXENE-CARBON ALLOTROPES-FOUNDED SUPERCAPACITORS (SCS)

The spray inoculation of Ti_3C_2Tx MXene on nickel-cobalt sulfide ($NiCO_2S_4$) at carbon cloth (CC) substratum caused $Ti_3C_2Tx/NiCO_2S_4$@CC composite (TNSC), which showed a gravimetric capacity of 2,326 F/g at 1 A/g and superior cyclic firmness of 93.8% at current compactness of 10 A/g [120]. The electrochemical presentation of the complex probe may be altered by varying the frame load of Ti_3C_2Tx MXene, which underwrites electrical conduction and brings about the penetrability of ions to nickel-cobalt sulfide. The partially solid-stage supple supercapacitor manufactured by amassing nitrogen-nobbled AC-covered carbon cloth (NACC) as the negative electrode and TNSC as the positive electrode in PVA/KOH gel medium displayed power compactness of 800 W/kg at an energy concentration of 57.5 Wh/kg, indicating a coulombic efficacy of 99.2% and a 90.2% capacity holding beyond 5,000 charge-discharge sequences (Figure 1.11).

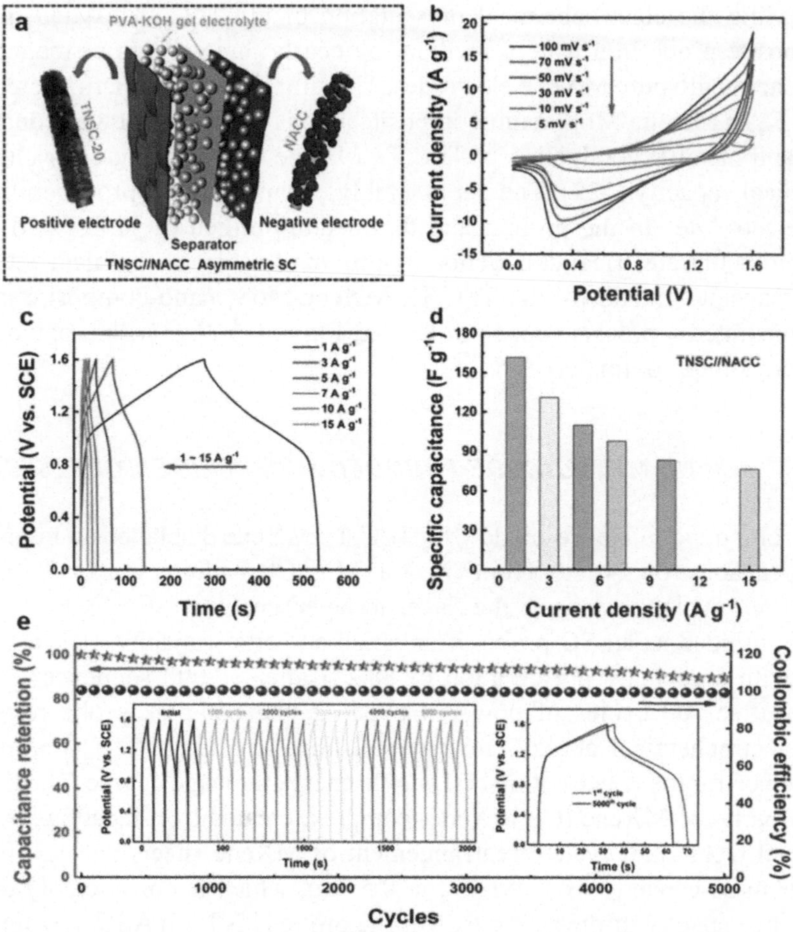

FIGURE 1.11 The electrochemical presentation of an asymmetric supercapacitor (ASC) comprising of NACC (positive) and TNSC (negative) electrodes in a PVA/KOH gel watery electrolyte is (a) the construction of ASC; (b) CV arcs at diverse probe rates; (c) and (d) GCD arcs and specific capacity at diverse current densities; and (e) capacity holding and coulombic efficacy of ASC.

Source: Reproduced with permission [120].

1.5.4 MXENE-TMD FOUNDED SUPERCAPACITORS (SCS)

Rakhi et al. [121] pointed out that the straight chemical creation of ε-MnO_2 whiskers on the surface of MXene leaves heightened the shallow area, thus increasing the electrochemical presentation of the Ti_3C_2Tx/ε-MnO_2 nano-composite probe in a watery medium by nearly three orders of magnitude compared with pure MXene electrodes. The simplistic hydrothermal creation of Ti_3C_2Tx MXene/MoS_2 nano-composite joint is the upshot of web construction and metallic conduction of Ti_3C_2Tx MXene to impart a greater electrochemical capacity [122]. The net assembly enhanced the approachability of ion electrolytes to the probes, and the metallic milieu of Ti_3C_2Tx MXene offered better electrical conduction, improving the electrochemical act. The supercapacitor made by the Ti_3C_2Tx MXene/MoS_2 nano-composite probe demonstrated a precise capacity of 115.2 F/g at 0.5 A/g, with a coulombic efficacy of approximately 90%.

1.5.5 MXENE-METAL OXIDE-FOUNDED SUPERCAPACITORS (SCS)

The compound of Co-Fe-oxide with Ti_3C_2Tx MXene displayed an improved bulk capacity of 2467.6 F/cm^3 in a 1 M LiCl medium for Co-Fe-oxide insertion, which enlarged the layer-to-layer spacing of MXene panes. When utilized as an SC probe, it unveiled an aerial capacity of 356.4 mF/cm^2 with a capacity preservation of 88.2% after 10,000 sequences [123]. Even after 100 cycles of powered twisting, the complex probe reserved the electrochemical energy stowage steadiness, making it an appropriate substance for the construction of a supple energy stowage device. The Co_3O_4 nano-particles-MXene (Co-MXene) complex, created by a facetious hydrothermal technique, evaded rearrangement of MXene sheets and testified a better mass capacity of 1,081 F/g at 0.5 A/g, which is considerably better than the capacity of the primeval MXene probe [124]. An ASC was formed using Co-MXene@Nickel foam as the positive electrode and PANI resultant carbon@carbon fiber paper (PANIC@CFP) as the negative electrode in a 6 M KOH solution, with a micropermeable polypropylene spacer between them. The ASC exhibited a higher voltage range of 1.4 V and a mass capacity of 95.71 F/g at 1 A/g. The ASC achieved a superior energy density of 26.06 Wh/kg at 700 W/kg and maintained 83% capacity after 8,000 cycles at 2 A/g.

1.5.6 MXENE-POLYMER FOUNDED SUPERCAPACITORS (SCS)

Even though MXenes have shown superior performance in various electrochemical energy storage strategies, the reconstruction of MXene layers hinders ion transfer. It has also been observed that the electrochemical performance of MXene-based probes decreases with increasing probe depth. These limitations of MXene probes can be overcome by constructing complexes of MXenes with polymers, which greatly enhance flexibility and electrochemical performance [125]. It has been revealed that mono-intercalated MXene layers exhibit better surface hydrophilicity and compatibility with polymers due to their high length-to-width ratio compared to multi-layered MXene layers [126]. Therefore, multilayered MXenes are delaminated prior to creating composites to enhance their electrochemical performance. Cheng et al. [127] fabricated a zinc hybrid micro-supercapacitor (ZHMSC) using a capacitor-type MXene/bacterial cellulose (BC)@polypyrrole (PPy) probe and a battery-type CNTs@MnO_2 probe in a 2 M $Zn(CF_3SO_3)_2$–0.1 M $MnSO_4$/PAM hydrogel. The ZHMSC shares the characteristics of both capacitor and battery-type probes. It demonstrates an aerial capacity of 290 mF/cm^2 at 0.75 mA/cm^2, a higher energy density of 145.4 μWh/cm^2, and a superior power density of 0.36 mW/cm^2 with 95.8% cycle stability over 25,000 cycles (Figure 1.12). The integration of conducting PPy polymer with pure MXene-based ZHMSC results in a remarkable increase in areal capacity, energy density, and power density. The authors also observed the retention of electrochemical performance during stretchable deformation up to 400%, making it a promising material for the fabrication of stretchable SCs. Vahidmohammadi et al. [128] investigated the Ti_3C_2Tx MXene compound, which was formed by in-situ polymerization of PANI on the surface of MXene. The compound exhibited a bulk capacity of 1,682 F/cm^3 and a mass capacity of 503 F/g. The researchers also observed that the composite electrode, with an areal density of 23.82 mg/cm^2 and a thickness of 90 μm, demonstrated a specific capacity of approximately 336 F/g, with 98.3% capacity retention after 10,000 cycles. The intercalation of PANI nanoparticles between the interlayers of MXene resulted in a pillaring effect, enhancing ion transport in MXene electrodes. This study paves the way for the development of thick electrodes with higher capacity.

Table 1.2 exemplifies the capacitive presentation of MXenes and a few related complex-based symmetric and asymmetric supercapacitors (ASCs) [129].

1.6 THEORETICAL INSIGHT FOR SUPERCAPACITOR

Many intricacies of the charge storage mechanism in electrochemical storage devices, which are difficult to comprehend physically, can be theoretically explained using quantum calculations such as molecular dynamics, Monte

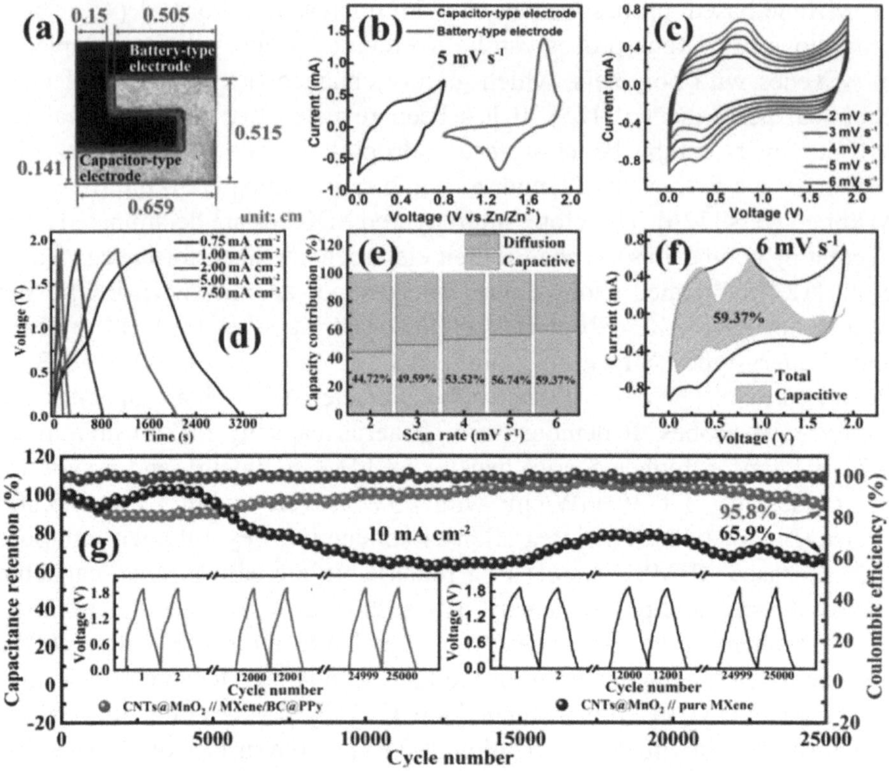

FIGURE 1.12 The electrochemical act of the zinc hybrid micro-supercapacitor with capacitor type MXene/bacterial cellulose (BC)@polypyrrole (PPy) probe and battery type CNTs@MnO_2 electrode; (a) strategy of ZHMSC; (b) CV arcs of capacitor type and battery type electrodes; (c) CV curve of ZHMSC at changeable probe rate; (d) GCD arc at variable current densities; (e) pseudocapacitive influence of capacitor type probe; (f) CV arc demonstrating 59.37% capacitive regulated involvement; and (g) cycle lifespan of ZHMSC.

Source: Reproduced with permission [127].

Carlo, and density functional theory (DFT). These electrochemical capacitors can store energy through charge distribution at the interface of electrode materials and electrolyte. Two types of ECs can be defined, depending on the charge storage process [139]. Generally, materials accumulate charge through the electrostatic adsorption of ions at the electrode-electrolyte interface, without involving any redox mechanism as illustrated in Figure 1.13(a) – (e). As h18, h49, and ISV all have a significant amount of charge, they are likely associated with lower capacitance. Also, Figure 1.13(d) exhibits the correlation between the probability density of average

TABLE 1.2 The Electrochemical Performance of MXene-based Symmetric and Asymmetric Supercapacitor Electrodes [129]

MXene Supercapacitor Electrode (Cathode/Anode)	Electrolyte	Capacitance	Energy Density	Power Density	Capacitance Retention	References
$Ti_3C_2T_x$ MXene//activated carbon	PVA/Na_2SO_4	7.8 mF/cm^2	3.5 mWh/cm^3	100 mW/cm^3	10,000 cycles 91.4%	[130]
ONCFT//MNCFT	PVA/H_2SO_4	780 mF/cm^2	277.3 µWh/cm^2	624 µW/cm^2	30,000 cycles 90%	[131]
$Ti_3C_2T_x$/CNF/PC	PVA/KOH	143 mF/cm^2 at 0.1 MA/cm^2	2.4 µ Wh/cm^2	17.5 µW/cm^2	10,000 cycles 90%	[132]
$NiCo_2$-LDHs@MXene/rGO//MXene/rGO	2 M KOH	240 F/g at 1 A/g	65.3 Wh/kg	700 W/kg	10,000 cycles 92.8%	[133]
$Ti_3C_2T_x$-based-Zn ion hybrid (0.851 µm)	PVA/$ZnCl_2$	72.02 mF/cm^2 and 662.53 F/cm^3	0.02 mWh/cm^2	0.50 mW/cm^2	50,000 cycles 80%	[134]
MXene/BC@PPy//CNTs@MnO_2	2 M $Zn(CF_3SO_3)_2$– 0.1 M $MnSO_4$/PAM hydrogel	290 mF/cm^2	145.4 µWh/cm^2	0.36 mW/cm^2	25,000 cycles 95.8%	[127]
MXene/RGo	FVA/H_3PO_4	372.2 mF/cm^2 and 586.4 F/cm^3 at 10 mV/s	13.03 mWh/cm^3	0.59 W/cm^3	3,000 cycles 100%	[114]
RGM-H_2SO_4	1 M H_2SO_4	454.9 F/cm^3 at 10 mV/s	39.4 Wh/L	180 W/L	–	[135]
V_2O_5//$Ti_3C_2T_x$ MXene (ZIHMSC)	$ZnSO_4$ dissolved PAM hydrogel.	129 mF/cm^2 at 0.34 mA/cm^2	48.9 µWh/cm^2	673 µW/cm^2	10,000 cycles 77%	[136]
Acid treated Ti_3C_2//RGO	PVA/H_2SO_4	256 F/cm^3 at 5 mV/s	58.4 mWh/cm^3 at 20.2 Wh/kg	7466.0 mW/cm^3 at 2583.0 W/kg	10,000 cycles 92.4%	[137]
MGH//PGH	1 M KCl	137.5 F/g at 10 mV/s	30.3 Wh/kg	1.13 kW/kg	10,000 cycles 81%	[138]

charge and the probability of highly charged atoms. In Figure 1.13(e), the simulation process of the electrodes (ZTC electrodes with graphene) within the z direction, where graphene caps the electrodes on both sides of the capacitor cell, has been discussed for FAU_1 (Faujasite). Molecular dynamics simulations are effective for investigating this type of framework based on redox reactions. However, molecular dynamics simulations usually include electrostatic and van der Waals contacts among molecules [139]. On the other hand, the charge storage mechanism is different for redox-type materials, known as pseudocapacitive materials. They accumulate the charge through a redox mechanism at the very close surface of the material. The analysis of the charge storage mechanism of a supercapacitor basically depends on its energy density. The energy density (E) and power density (P) can be defined as:

$$E = \frac{CXV^2}{2} and P = \frac{V^2}{4R} \tag{3}$$

where 'C,' 'V,' and 'R' represent capacitance, voltage, and resistance values, respectively. To achieve high energy density, the capacitance and operating potential values need to be maximized.

1.6.1 QUANTUM CAPACITANCE

Quantum capacitance is well-defined as the disparity of electrical charge q commensurating the change of electrochemical potential $\bar{\mu}$, i.e., $C_{\bar{\mu}} = \frac{dq}{d\bar{\mu}}$. In the case of a parallel-plate capacitor where one or both of the plates have a low density of states, the capacitance is not obtained through the usual formulation for parallel-plate capacitors, C_e. Instead, an alternative capacitance, C_q, is introduced as if there were an additional capacitor in series. The capacitance associated with the density of states of the plates is known as the quantum capacitance (C_q). The corresponding capacitance is referred to as electrochemical capacitance. $\frac{1}{C_{\bar{\mu}}} = \frac{1}{C_e} + \frac{1}{C_q}$ It is particularly significant for low-density-of-states schemes, such as a layered electronic scheme in a semiconductor surface, interface, or graphene, and may be utilized to conceptualize a pragmatic energy functional of electron density [140].

When a voltmeter is used to measure an electronic device, it doesn't actually measure the pure electric potential (also known as Galvani potential). Instead, it measures the electrochemical potential, known as the "Fermi

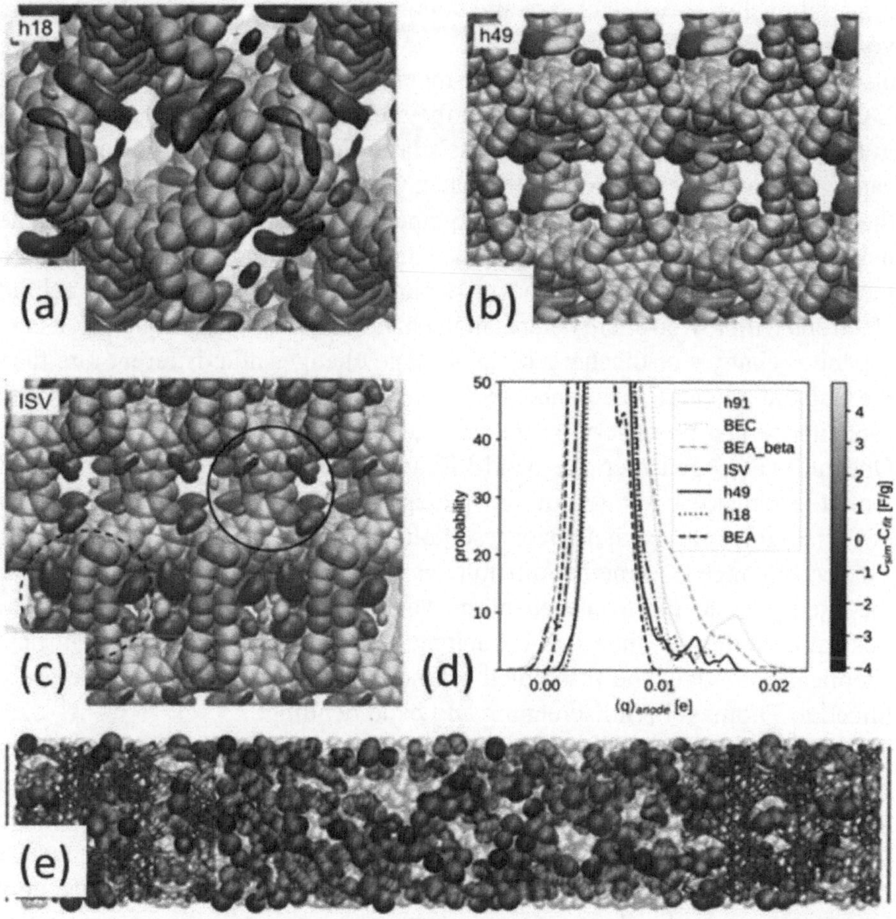

FIGURE 1.13 (a)–(c) Inspection of atomic charge different h18, h49, and ISV materials simulated for anion; (d) average charge distribution (probability density) in anode; and (e) simulation geometry of Faujasite (FAU) (scaling of color from 0 (white) to 0.01 (red)), total charge indicated by purple color with cut of greater than 0.01e and green show total charge less than 0.1e.

Source: Reproduced with permission [139].

level difference," which is the total free energy change per electron, taking into account not only its electric potential energy but also all other forces and effects on the electron (e.g., the kinetic energy in its wavefunction). For example, in a p-n junction at equilibrium, there is a Galvani potential (built-in potential) across the junction, but the "voltage" across it is zero (in the sense that a voltmeter would measure zero voltage). In a capacitor, there is a relationship between charge and voltage, Q=CV. As explained above, the voltage can be divided into two parts: the Galvani potential, and everything else. In a traditional metal-insulator-metal capacitor, the Galvani potential is the only relevant input. Consequently, the capacitance may be considered in a straightforward way using Gauss's law. However, if one or both of the capacitor plates are semiconductors, then Galvani potential is not necessarily the sole significant influence on capacitance. As the capacitor charges up, the negative plate fills up with electrons, which occupy higher-energy states in the band structure, while the positive plate loses electrons, leaving behind electrons with lower-energy states in the band structure. Consequently, as the capacitor charges or discharges, the voltage changes at a different rate than the Galvani potential difference. In these circumstances, one cannot compute capacitance just by observing the general geometry and using Gauss's law. One must also consider the band-filling/band-emptying effect, associated with the density of states of the plates. The band-filling/band-emptying effect modifies the capacitance, resembling a second capacitance in series. This capacitance is termed as quantum capacitance since it is associated with the energy of an electron's quantum wavefunction. Some experts refer to this same notion as chemical capacitance, as it is related to the electrons' chemical potential. The principles behind quantum capacitance are closely linked to Thomas–Fermi screening and band bending.

The quantum capacitance of graphene is pertinent to consider and model gated graphene [141]. It is also applicable for carbon nanotubes [142]. In the research of dye-sensitized solar cells, the quantum capacitance of the sintered TiO_2 nanoparticle probe is a vital result [143]. Quantum capacitance can be significant in capacitance-voltage profiling. When supercapacitors are researched in detail, quantum capacitance plays a significant part [144]. Theoretically, it is computed from the density of states using the relation [145]:

$$C_Q = e^2 \int_{-\infty}^{\infty} D(E) F_T(E - e\varphi_G) dE \tag{4}$$

where D(E) and ϕ_G are the density of states and electrode potential of the system under consideration, respectively, while FT(E) is the thermal broadening function due to thermal effect expressed as:

$$F_T(E) = (4K_BT)^{-1}\text{sech}^2(E/2k_BT) \tag{5}$$

Figure 1.14 compares the quantum capacitance between pristine VS_4 and VSe_2, as well as their hybrids when coupled with MXene, as a function of electrode potential [146, 147]. It is evident that hybridization significantly enhances the specific capacitance in both cases.

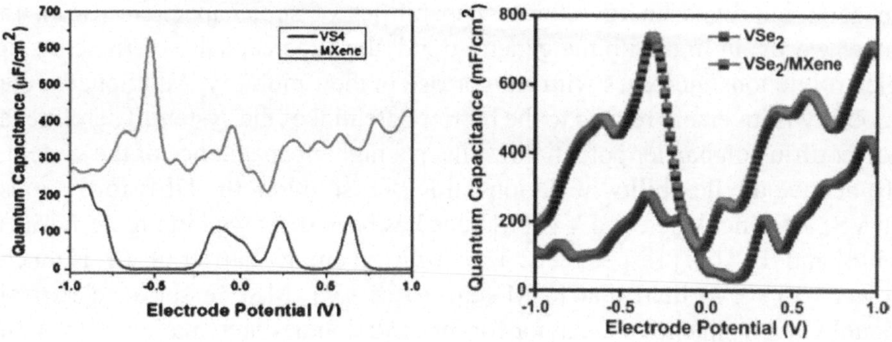

FIGURE 1.14 Variation of QC as a function of electrode potential for pristine (a) VS_4; and (b) VSe_2 and their respective hybrids with MXene.

Source: Reproduced with permission [146, 147].

1.6.2 DIFFUSION ENERGY BARRIER

The energy barrier is the difference between the energy of the diffusing atom in the transition complex and the energy in the kink site position. High-rate electrode materials are crucial for enabling rapid charging of EES devices [148]. The most commonly employed technique to enhance rate capability is the fabrication of nanosized materials [149], which exhibit short diffusion distances and a large interaction area with the electrolyte. Unfortunately, the low tap density of these nanomaterials severely limits the achievable volumetric energy density of the strategies. Obviously, micron-sized materials are more favorable for practical energy storage applications. However, micron-sized particles often exhibit lower rate capability due to the significant diffusion barrier caused by a long diffusion path and separation of the inner portion of the particles from the electrolyte. Since rate capability is mainly limited by the slow dispersion of solid ions in the bulk and restricted ion transport at the interface between the active material and electrolyte, rapid dispersion kinetics within the lattice and efficient ion transport at the interface are highly desired to reduce the diffusion barrier and improve the rate performance of micron-sized materials. Theoretically, it is possible to calculate the diffusion energy barrier (DEB) for the electrolyte ions, and a lower barrier indicates an improvement in rate capability.

VS_4 exhibits redox-type capacitive behavior, where charge transporters are generated. The movement of these charge carriers plays a significant role in the performance of charge storage. The higher mobility of ions in electrolytes leads to improved energy storage capabilities of supercapacitors (SCs). In other words, in line with the general trend, the supercapacitive efficiency of electrolyte ions increases with an increase in their mobility. Additionally, the mobility is inversely related to the barrier potential of the system. Therefore, a lower diffusion barrier potential results in a higher capacitance of the system. To deduce the flexibility of K^+ ions in K_2SO_4 solution, the DEB for K^+ ions in VS_4, MXene, VSe_2, and VSe_2/MXene has been displayed in Figure 1.15(a) [146] and (b) [147] respectively. The diffusion energy barrier in VS_4 is much lower (0.17 eV) than that in MXene (0.28 eV). MXene shows electrical double-layer capacitive behavior for intercalation/de-intercalation of ions. In the case of VSe_2, the MXene hybrid appears to offer a lower DEB, indicating higher mobility and a better charge transfer rate.

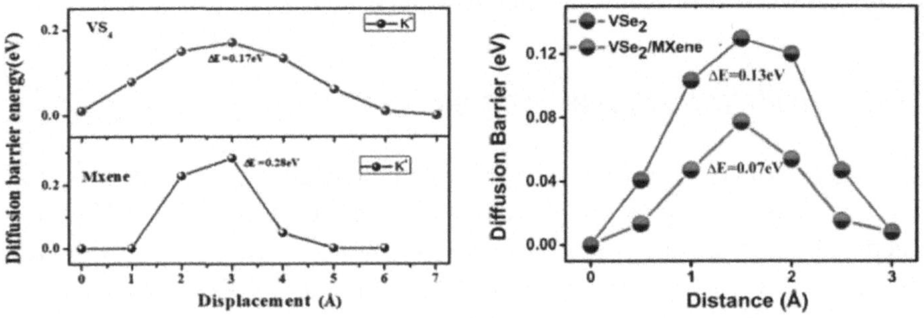

(a) (b)

FIGURE 1.15 Diffusion energy barrier for the K atom to travel between equivalent positions in (a) pristine VS_4 and MXene; and (b) pristine VSe_2 and VSe_2-MXene hybrid.

Source: Reproduced with permission [146, 147].

1.6.3 INDUCED VOLTAGE

The amount of current flowing through a capacitor depends on both the capacitance and the rate at which the voltage is changing. If the voltage across a capacitor increases rapidly, a large positive current will flow through the capacitor. A change in the current flowing through a conductor induces a voltage in that conductor, as well as in all nearby conductors. The induced voltage opposes the change in current that caused it. Inductance is a consequence of two principles of physics. Firstly, a steady current flowing through a conductor creates a persistent magnetic field. Secondly, a changing magnetic field induces a voltage in all nearby conductors, including the conductor that was used to create the magnetic field in the first place. When these two laws are combined, the resulting effect is inductance. Just as resistors are included to introduce an expected resistance in a circuit, and capacitors are used to provide a desired capacitance, inductors are electrical components used to introduce a desired amount of inductance into the circuit.

For theoretical calculation of induced voltage, involving supercapacitor, V is defined as [150]:

$$V = -\frac{E_{K_{x1}} - (E_{K_{x2}} + (x_1 - x_2)E_K)}{(x_1 - x_2)e} \tag{6}$$

where; X_1, X_2 are the concentrations of K^+ ions; and $E_{K_{x1}}$ is the energy of the system per formula unit for concentration X_1. For example, Pathak et al. [151] computed the average voltage of $NiCo_2O_4$@MWCNT composite with three different K^+ ion concentrations, i.e., 0.625, 0.25, and 0.5, using the above equation. The average voltage of the system for various K^+ ion concentrations is shown in Table 1.3, which is similar to the predicted values for lithium and sodium ions [30].

TABLE 1.3 The Average Voltage of the K^+ Ion on $NiCo_2O_4$ for Different K^+ Ion Concentrations*

$E_{K^+_{x1}NiCo_2O_4}$ (eV)	$E_{K^+_{x2}NiCo_2O_4}$ (eV)	Voltage (V)
–44.911	–45.398	1.38
–44.853	–45.398	2.08
–44.853	–44.911	3.01

**Where; 'x' represents the concentrations of the* K^+ *ion* (*x*: 0.625, 0.25, and 0.5); and $E_{K^+_{x1}NiCo_2O_4}$ is the total energy of the system per formula unit [151].

1.7 CONCLUSION AND FUTURE DIRECTIONS

In this chapter, a detailed overview of the supercapacitor is provided, including the role of morphology, nano-materials that can be used as electrolyte materials, and theoretical factors that affect the performance of SCs. SCs, as a new storage device, are highly attractive in the field of energy storage and harvesting. Previous research on SCs has shown effective ways to achieve new heights in this field, but there are still limitations that provide new directions for scientists and research communities to overcome these challenges and improve the energy density performance. Similarly, our future task is to find the ideal parameters to model a supercapacitor for military applications. By increasing the low rated voltage, we can eliminate the need for any series combination of SCs for practical applications. The ultimate focus of researchers in industry and research is to achieve high power density in SCs. This objective can be achieved not only by making SCs, but also by finding electroactive NMs with high electrochemical performance.

KEYWORDS

- **2D nanostructures**
- **cycle life**
- **electrochemical properties**
- **energy density**
- **energy storage**
- **metal dichalcogenides**
- **power density**
- **supercapacitor**

REFERENCES

1. Tanaka, N., (2010). *World Energy Outlook 2010.* International Energy Agency. Paris: IEA.
2. *World Gross Electricity Production by Source, 2019*–Charts–Data & Statistics. International Energy Agency.
3. Powell, C. A., & Bryan, D. M., (2008). Materials challenges in advanced coal conversion technologies. *MRS Bulletin, 33*(4), 309–315.
4. Arunachalam, V. S., & Fleischer, E. L., (2008). The global energy landscape and materials innovation. *MRS Bulletin, 33*(4), 264–288.
5. Meier, P. J., Paul, P. H. W., Gerald, L. K., & Paul, L. D., (2005). US electric industry response to carbon constraint: A life-cycle assessment of supply side alternatives. *Energy Policy, 33*(9), 1099–1108.
6. Ginley, D., Martin, A. G., & Reuben, C., (2008). Solar energy conversion toward 1 terawatt. *MRS Bulletin, 33*(4), 355–364.
7. Holdren, J. P., (2007). Energy and sustainability. *Science, 315*(5813), 737.
8. Lee, J. C. Y., & Jason, F. M., (2021). An overview of wind-energy-production prediction bias, losses, and uncertainties. *Wind Energy Science, 6*(2), 311–365.
9. Kintner-Meyer, M. C. W., Yang, Z., Zhang, J., Lu, X., Choi, D., Lemmon, J. P., & Liu, J., (2011). Electrochemical energy storage for green grid. *Chem. Rev., 111*, 3577–3613.

10. Lee, B. S., & David, E. G., (2008). Electricity storage: The Achilles' heel of renewable energy. *Chemical Engineering Progress, 104*(3), S29–S32.
11. Burke, A., (2007). R&D considerations for the performance and application of electrochemical capacitors. *Electrochimica Acta, 53*(3), 1083–1091.
12. Pomerantseva, E., Francesco, B., Xinliang, F., Yi, C., & Yury, G., (2019). Energy storage: The future enabled by nanomaterials. *Science, 366*(6468), eaan8285.
13. Chen, H., Thang, N. C., Wei, Y., Chunqing, T., Yongliang, L., & Yulong, D., (2009). Progress in electrical energy storage system: A critical review. *Progress in Natural Science, 19*(3), 291–312.
14. Aneke, M., & Meihong, W., (2016). Energy storage technologies and real-life applications–A state of the art review. *Applied Energy, 179*, 350–377.
15. Gür, T. M., (2018). Review of electrical energy storage technologies, materials, and systems: Challenges and prospects for large-scale grid storage. *Energy & Environmental Science, 11*(10), 2696–2767.
16. Díaz-González, F., Andreas, S., Gomis-Bellmunt, O., & Villafáfila-Robles, R., (2012). A review of energy storage technologies for wind power applications. *Renewable and Sustainable Energy Reviews, 16*(4), 2154–2171.
17. Liu, C., Feng, L., Lai-Peng, M., & Hui-Ming, C., (2010). Advanced materials for energy storage. *Advanced Materials, 22*(8), E28–E62.
18. Long, J. W., Bruce, D., Debra, R. R., & Henry, S. W., (2004). Three-dimensional battery architectures. *Chemical Reviews, 104*(10), 4463–4492.
19. Janek, J., & Wolfgang, G. Z., (2016). A solid future for battery development. *Nature Energy, 1*(9), 1–4.
20. Van, N. R., (2014). A better battery. *Nature, 507*(7490), 26.
21. Raza, W., Faizan, A., Nadeem, R., Yiwei, L., Ki-Hyun, K., Jianhua, Y., Sandeep, K., et al., (2018). Recent advancements in supercapacitor technology. *Nano Energy, 52*, 441–473.
22. Zhang, L. L., & Zhao, X. S., (2009). Carbon-based materials as supercapacitor electrodes. *Chemical Society Reviews, 38*(9), 2520–2531.
23. Snook, G. A., Pon, K., & Adam, S. B., (2011). Conducting-polymer-based supercapacitor devices and electrodes. *Journal of Power Sources, 196*(1), 1–12.
24. Muzaffar, A., Basheer, A. M., Kalim, D., & Jagannathan, T., (2019). A review on recent advances in hybrid supercapacitors: Design, fabrication, and applications. *Renewable and Sustainable Energy Reviews, 101*, 123–145.
25. Salanne, M., Benjamin, R., Katsuhiko, N., Katsumi, K., Taberna, P. L., Clare, P. G., Bruce, D., & Patrice, S., (2016). Efficient storage mechanisms for building better supercapacitors. *Nature Energy, 1*(6), 1–10.
26. Patra, A., Namsheer, K., Jeena, R. J., Surjit, S., Brahmananda, C., & Chandra, S. R., (2021). Understanding the charge storage mechanism of supercapacitors: In situ/operand spectroscopic approaches and theoretical investigations. *Journal of Materials Chemistry A 9*(46), 25852–25891.
27. Sahoo, S., Pravin, W., Saroj, K. N., Chandra, S. R., & Dattatray, J. L., (2021). Recent developments in self-powered smart chemical sensors for wearable electronics. *Nano Research, 14*(11), 3669–3689.
28. Yu, X., Sol, Y., Jeong, S. Y., Pallab, B., Libin, W., Seung, W. L., Xianluo, H., & Ho Seok, P., (2018). Emergent pseudocapacitance of 2D nanomaterials. *Advanced Energy Materials, 8*(13), 1702930.

29. Jiang, Y., & Jinping, L., (2019). Definitions of pseudocapacitive materials: A brief review. *Energy & Environmental Materials, 2*(1), 30–37.
30. Fan, S., Xiaolong, Z., Hongda, D., Lin, G., Chengjun, X., Wei, L., Yan-Bing, H., et al., (2017). Theoretical investigation of the intercalation chemistry of lithium/sodium ions in transition metal dichalcogenides. *The Journal of Physical Chemistry C, 121*(25), 13599–13605.
31. Isacfranklin, M., Yuvakkumar, R., Senthil, K. P., Thirumal, V., Ravi, G., & Dhayalan, V., (2021). Hydrogen free direct growth carbon nanorod as a promising electrode in symmetric supercapacitor applications. *Progress in Organic Coatings, 158*, 106379.
32. Mane, V. J., Kale, S. B., Ubale, S. B., Lokhande, V. C., & Lokhande, C. D., (2021). Enhanced specific energy of silver-doped MnO_2/graphene oxide electrodes as facile fabrication symmetric supercapacitor device. *Materials Today Chemistry, 20*, 100473.
33. Beknalkar, S. A., Teli, A. M., Harale, N. S., Shin, J. C., & Patil, P. S., (2021). Construction of IrO_2@ Mn_3O_4 core-shell heterostructured nanocomposites for high performance symmetric supercapacitor device. *Journal of Alloys and Compounds, 887*, 161328.
34. Mathew, M., & Chandra, S. R., (2021). Electrochemical biosensors based on Ti_3C_2Tx MXene: Future perspectives for on-site analysis. *Current Opinion in Electrochemistry, 30*, 100782.
35. Wu, Y., Hao, C., Yingzhuo, L., Jin, Y., Xinqiang, Z., Yu, Z., Gaobo, L., et al., (2021). Rational design of cobalt–nickel double hydroxides for flexible asymmetric supercapacitor with improved electrochemical performance. *Journal of Colloid and Interface Science, 581*, 455–464.
36. Ghosh, K., & Martin, P., (2021). MXene and MoS_{3-x} coated 3D-printed hybrid electrode for solid-state asymmetric supercapacitor. *Small Methods, 5*(8), 2100451.
37. Shao, Y., El-Kady, M. F., Jingyu, S., Yaogang, L., Qinghong, Z., Meifang, Z., Hongzhi, W., et al., (2018). Design and mechanisms of asymmetric supercapacitors. *Chemical Reviews, 118*(18), 9233–9280.
38. Brousse, T., Daniel, B., & Jeffrey, W. L., (2015). To be or not to be pseudocapacitive? *Journal of The Electrochemical Society, 162*(5), A5185.
39. Tan, H. T., Wenping, S., Libo, W., & Qingyu, Y., (2016). 2D transition metal oxides/hydroxides for energy-storage applications. *ChemNanoMat, 2*(7), 562–577.
40. Moraes, L. P. R., Jun, M., Fabio, C. F., & Ziqi, S., (2020). Two-dimensional metal oxide nanomaterials for sustainable energy applications. In: *2D Nanomaterials for Energy Applications* (pp. 39–72). Elsevier.
41. Jin, H., Chunxian, G., Xin, L., Jinlong, L., Anthony, V., Yan, J., Yao, Z., & Shi-Zhang, Q., (2018). Emerging two-dimensional nanomaterials for electrocatalysis. *Chemical Reviews, 118*(13), 6337–6408.
42. Lee, S., Xiaoyan, J., In Young, K., Tae-Ha, G., Ji-Won, C., Sahn, N., & Seong-Ju, H., (2016). Superior additive of exfoliated RuO_2 nanosheet for optimizing the electrode performance of metal oxide over graphene. *The Journal of Physical Chemistry C, 120*(22), 11786–11796.
43. Xia, A., Wanru, Y., Jue, Y., Guoqiang, T., Huijun, R., & Chun, L., (2019). Synthesis of porous δ-MnO_2 nanosheets and their supercapacitor performance. *Journal of Electro analytical Chemistry, 839*, 25–31.
44. Feng, L., Yufu, Z., Hongyan, D., &Chaoying, N., (2014). Recent progress in nickel-based materials for high performance pseudocapacitor electrodes. *Journal of Power Sources, 267*, 430–444.

45. Low, W. H., Poi, S. K., Siew, S. L., Chiu, W. S., & Ejikeme, R. E., (2019). Recent development of mixed transition metal oxide and graphene/mixed transition metal oxide-based hybrid nanostructures for advanced supercapacitors. *Journal of Alloys and Compounds, 775*, 1324–1356.
46. Lu, Q., Jingguang, G. C., & John, Q. X., (2013). Nanostructured electrodes for high-performance pseudocapacitors. *Angewandte Chemie International Edition, 52*(7), 1882–1889.
47. Zheng, J. P., & Jow, T. R., (1995). A new charge storage mechanism for electrochemical capacitors. *Journal of the Electrochemical Society, 142*(1), L6.
48. Yu, Z., Laurene, T., Lei, Z., & Jayan, T., (2015). Supercapacitor electrode materials: Nanostructures from 0 to 3 dimensions. *Energy & Environmental Science, 8*(3), 702–730.
49. Wang, G., Lei, Z., & Jiujun, Z., (2012). A review of electrode materials for electrochemical supercapacitors. *Chemical Society Reviews, 41*(2), 797–828.
50. Al-Rubaye, S., Ranjusha, R., Shi, X. D., & Zhenxiang, C., (2017). Facile synthesis of a reduced graphene oxide wrapped porous $NiCo_2O_4$ composite with superior performance as an electrode material for supercapacitors. *Journal of Materials Chemistry A, 5*(36), 18989–18997.
51. Patil, S. S., Deepak, P. D., Virendrakumar, G. D., Mohaseen, S. T., Jalindar, D. A., Gomez-Romero, P., Sanjay, S. K., et al., (2016). Fern-like rGO/BiVO4 hybrid nanostructures for high-energy symmetric supercapacitor. *ACS Applied Materials & Interfaces, 8*(46), 31602–31610.
52. Zhi, M., Chengcheng, X., Jiangtian, L., Ming, L., & Nianqiang, W., (2013). Nanostructured carbon–metal oxide composite electrodes for supercapacitors: A review. *Nanoscale, 5*(1), 72–88.
53. Kandasamy, M., Surjit, S., Saroj, K. N., Brahmananda, C., & Chandra, S. R., (2021). Recent advances in engineered metal oxide nanostructures for supercapacitor applications: Experimental and theoretical aspects. *Journal of Materials Chemistry A, 9*(33), 17643–17700.
54. Anasori, B., Majid, B., & Yury, G., (2014). Graphene–transition metal oxide hybrid materials. *Mater. Today, 17*(5), 253–254.
55. Zhong-Shuai, W., Guangmin, Z., Li-Chang, Y., Wencai, R., Feng, L., & Hui-Ming, C., (2012). Graphene/metal oxide composite electrode materials for energy storage. *Nano Energy, 1*(1), 107–131.
56. Zhang, J., Jianwen, J., & Zhao, X. S., (2011). Synthesis and capacitive properties of manganese oxide nanosheets dispersed on functionalized graphene sheets. *The Journal of Physical Chemistry C, 115*(14), 6448–6454.
57. Ge, C., Zhaohui, H., Binhong, H., Fanyan, Z., Jianguo, C., Yiming, L., & Yafei, K., (2012). Three-dimensional flower-like nickel oxide supported on graphene sheets as electrode material for supercapacitors. *Journal of Sol-Gel Science and Technology, 63*(1), 146–152.
58. Peng, L., Xu, P., Borui, L., Changzheng, W., Yi, X., & Guihua, Y., (2013). Ultrathin two-dimensional MnO_2/graphene hybrid nanostructures for high-performance, flexible planar supercapacitors. *Nano Letters, 13*(5), 2151–2157.
59. Tan, H. Y., Bao, Z. Y., Lin, L. C., Tao, C., Xin, L. Z., Xing, H. L., Wei, L. L., & Zhao, Y. R., (2017). Layer-dependent growth of two-dimensional Co_3O_4 nanostructure arrays on graphene for high performance supercapacitors. *Journal of Alloys and Compounds, 696*, 1180–1188.

60. Shrestha, K. R., Syam, K., Nam, H. K., & Joong, H. L., (2019). A spinel $MnCo_2O_4$/NG 2D/2D hybrid nanoarchitectures as advanced electrode material for high performance hybrid supercapacitors. *Journal of Alloys and Compounds, 771*, 810–820.
61. Pan, H., Jianyi, L., & YuanPing, F., (2010). Carbon nanotubes for supercapacitor. *Nanoscale Research Letters, 5*(3), 654–668.
62. Zhou, H., Lusi, Z., Dongyang, Z., Shuangqiang, C., Paul, R. C., Xiong, H., Mike, C., et al., (2016). A universal synthetic route to carbon nanotube/transition metal oxide nano-composites for lithium ion batteries and electrochemical capacitors. *Scientific Reports, 6*(1), 1–11.
63. Mondal, A. K., Hao, L., Zhen-Fang, L., &Guoxiu, W., (2016). Multiwall carbon nanotube-nickel cobalt oxide hybrid structure as high-performance electrodes for supercapacitors and lithium-ion batteries. *Electrochimica Acta, 190*, 346–353.
64. Barbieri, O., Hahn, M., Herzog, A., & Kötz, R., (2005). Capacitance limits of high surface area activated carbons for double layer capacitors. *Carbon, 43*(6), 1303–1310.
65. Zhang, X., Jiangshui, L., Pengyi, T., Xiaoliang, Y., Xinxing, P., Haolin, T., Shi-Gang, S., & Jan, F., (2017). A universal strategy for metal oxide anchored and binder-free carbon matrix electrode: A supercapacitor case with superior rate performance and high mass loading. *Nano Energy, 31*, 311–321.
66. Ramya, R., Sivasubramanian, R., & Sangaranarayanan, M. V., (2013). Conducting polymers-based electrochemical supercapacitors—Progress and prospects. *Electrochimica Acta, 101*, 109–129.
67. Li, S., Dongqing, W., Chong, C., Jinzuan, W., Fan, Z., Yuezeng, S., & Xinliang, F., (2013). Polyaniline-coupled multifunctional 2D metal oxide/hydroxide graphene nanohybrids. *Angewandte Chemie International Edition, 52*(46), 12105–12109.
68. Kuang, M., Xiao, Y. L., Fan, D., & Yu Xin, Z., (2015). Tunable design of layered $CuCo_2O_4$ nanosheets@ MnO_2 nanoflakes core–shell arrays on Ni foam for high-performance supercapacitors. *Journal of Materials Chemistry A, 3*(43), 21528–21536.
69. Liu, J., Jian, J., Chuanwei, C., Hongxing, L., Jixuan, Z., Hao, G., & Hong, J. F., (2011). Co_3O_4 nanowire @ MnO_2 ultrathin nanosheet core/shell arrays: A new class of high-performance pseudocapacitive materials. *Advanced Materials, 23*(18), 2076–2081.
70. Xu, K., Shan, M., Yuenian, S., Qilong, R., Jianmao, Y., Xiao, C., & Junqing, H., (2019). $CuCo_2O_4$ nanowire arrays wrapped in metal oxide nanosheets as hierarchical multicomponent electrodes for supercapacitors. *Chemical Engineering Journal, 369*, 363–369.
71. Cheng, Q., Jie, T., Jun, M., Han, Z., Norio, S., & Lu-Chang, Q., (2011). Graphene and nanostructured MnO_2 composite electrodes for supercapacitors. *Carbon, 49*(9), 2917–2925.
72. Soon, J. M., & Kian, P. L., (2007). Electrochemical double-layer capacitance of MoS_2nanowall films. *Electrochemical and Solid-State Letters, 10*(11), A250.
73. Cao, L., Shubin, Y., Wei, G., Zheng, L., Yongji, G., Lulu, M., Gang, S., et al., (2013). Direct laser-patterned micro-supercapacitors from paintable MoS_2 films. *Small, 9*(17), 2905–2910.
74. Chhowalla, M., Hyeon, S. S., Goki, E., Lain-Jong, L., Kian, P. L., & Hua, Z., (2013). The chemistry of two-dimensional layered transition metal dichalcogenides nanosheets. *Nature Chemistry, 5*(4), 263–275.
75. Voiry, D., Anandarup, G., Rajesh, K., Daniel, K., Takeshi, F., Mingwei, C., Tewodros, A., & Manish, C., (2015). Covalent functionalization of monolayered transition metal dichalcogenides by phase engineering. *Nature Chemistry,7*(1), 45–49.

76. Tang, Q., & De-en, J., (2015). Stabilization and band-gap tuning of the 1T-MoS_2 monolayer by covalent functionalization. *Chemistry of Materials, 27*(10), 3743–3748.
77. Acerce, M., Damien, V., & Manish, C., (2015). Metallic 1T phase MoS_2 nanosheets as supercapacitor electrode materials. *Nature Nanotechnology, 10*(4), 313–318.
78. Yin, Y., Yumin, Z., Tangling, G., Tai, Y., Xinghong, Z., Jiecai, H., Xianjie, W., et al., (2017). Synergistic phase and disorder engineering in 1T-$MoSe_2$ nanosheets for enhanced hydrogen-evolution reaction. *Advanced Materials, 29*(28), 1700311.
79. Piao, M., Jin, C., Xiao, W., Yao, C., Heng, Z., Chaolong, L., Haofei, S., & Min-Kyu, J., (2017). Hydrothermal synthesis of stable metallic 1T phase WS_2 nanosheets for thermoelectric application. *Nanotechnology, 29*(2), 025705.
80. Ambrosi, A., Zdeněk, S., & Martin, P., (2015). 2H→ 1T phase transition and hydrogen evolution activity of MoS_2, $MoSe_2$, WS_2 and WSe_2 strongly depends on the MX_2 composition. *Chemical Communications, 51*(40), 8450–8453.
81. Chen, H., Jianjun, J., Li, Z., Houzhao, W., Tong, Q., & Dandan, X., (2013). Highly conductive $NiCo_2S_4$ urchin-like nanostructures for high-rate pseudocapacitors. *Nanoscale, 5*(19), 8879–8883.
82. Xia, C., Li, P., Gandi, A. N., Schwingenschlögl, U., & Alshareef, H. N., (2015). Is $NiCo_2S_4$ really a semiconductor? *Chemistry of Materials, 27*(19), 6482–6485.
83. Bouchard, R. J., Russo, P. A., & Wold, A., (1965). Preparation and electrical properties of some thiospinels. *Inorganic Chemistry, 4*(5), 685–688.
84. El Radaf, I. M., Fouad, S. S., Ismail, A. M., & Sakr, G. B., (2018). Influence of spray time on the optical and electrical properties of $CoNi_2S_4$ thin films. *Materials Research Express, 5*(4), 046406.
85. Jae-Jin, S., (2015). *In situ* growth of hierarchical mesoporous $NiCo_2S_4$@ MnO_2 arrays on nickel foam for high-performance supercapacitors. *Electrochimica Acta, 166*, 302–309.
86. Mei, L., Ting, Y., Cheng, X., Ming, Z., Libao, C., Qiuhong, L., & Taihong, W., (2014). Hierarchical mushroom-like $CoNi_2S_4$ arrays as a novel electrode material for supercapacitors. *Nano Energy, 3*, 36–45.
87. Peng, T., Zhongyu, Q., Jun, W., Dalei, S., Jinyuan, L., Qi, L., & Peng, W., (2014). Construction of mass-controllable mesoporous $NiCo_2S_4$ electrodes for high performance supercapacitors. *Journal of Materials Chemistry A, 2*(45), 19376–19382.
88. Li, Z., Dian, Z., Chunyang, X., Jiqiang, N., Yijun, Z., Ziyang, Z., Yongjiang, W., & Yong, H., (2018). Reduced $CoNi_2S_4$ nanosheets with enhanced conductivity for high-performance supercapacitors. *Electrochimica Acta, 278*, 33–41.
89. Hu, W., Ruqi, C., Wei, X., Lilan, Z., Ni, Q., & Dinghua, B., (2014). $CoNi_2S_4$ nanosheet arrays supported on nickel foams with ultrahigh capacitance for aqueous asymmetric supercapacitor applications. *ACS Applied Materials & Interfaces,6*(21), 19318–19326.
90. Lin, L., Wen, L., Shaowei, Z., Yuqing, L., Gordon, G. W., & Jun, C., (2019). Two-dimensional transition metal dichalcogenides in supercapacitors and secondary batteries. *Energy Storage Materials, 19*, 408–423.
91. Muller, G. A., John, B. C., Hyung-Seok, K., Sarah, H. T., & Bruce, D., (2015). High performance pseudocapacitor based on 2D layered metal chalcogenide nanocrystals. *Nano Letters, 15*(3), 1911–1917.
92. Chao-Chi, T., Lu-Yin, L., Bing-Chang, X., & Yu-Shiang, C., (2016). Highly efficient supercapacitor electrode with two-dimensional tungsten disulfide and reduced graphene oxide hybrid nanosheets. *Journal of Power Sources, 320*, 78–85.

93. Shen, J., Jingjie, W., Liyuan, P., Marco-Tulio, F. R., ZhuQing, Z., Fangfang, Z., Xiang, Z., et al., (2016). $CoNi_2S_4$-Graphene-2D-$MoSe_2$ as an advanced electrode material for supercapacitors. *Advanced Energy Materials, 6*(13), 1600341.
94. Lin, L., Naihua, M., Yan, W., Shaowei, Z., Philippe, G., Zhimei, S., & Dan, A. A., (2016). Sulfur-depleted monolayered molybdenum disulfide nanocrystals for super electrochemical hydrogen evolution reaction. *ACS Nano, 10*(9), 8929–8937.
95. Lin, L., Naihua, M., Juntong, H., Shaowei, Z., Yanqiu, Z., David, D. H., Philippe, G., et al., (2017). A photocatalyst of Sulphur depleted monolayered molybdenum sulfide nanocrystals for dye degradation and hydrogen evolution reaction. *Nano Energy, 38*, 544–552.
96. Lin, L., Yaoxian, X., Shaowei, Z., Ian, M. R., Albert, C. M. O., & Dan, A. A., (2013). Fabrication of luminescent monolayered tungsten dichalcogenides quantum dots with giant spin-valley coupling. *ACS Nano, 7*(9), 8214–8223.
97. Tan, C., Zhimin, L., Apoorva, C., Yongqing, C., Yonghua, D., Yue, G., Ying, H., et al., (2018). Preparation of high-percentage 1T-phase transition metal dichalcogenide nanodots for electrochemical hydrogen evolution. *Advanced Materials, 30*(9), 1705509.
98. Li, H., Roland, Y. T., Siu, H. T., Xu, Z., & Edwin, H. T. T., (2015). Controllable synthesis of highly luminescent boron nitride quantum dots. *Small, 11*(48), 6491–6499.
99. Xu, S., Dian, L., & Peiyi, W., (2015). One-pot, facile, and versatile synthesis of monolayer MoS_2/WS_2 quantum dots as bioimaging probes and efficient electrocatalysts for hydrogen evolution reaction. *Advanced Functional Materials, 25*(7), 1127–1136.
100. Jeong, S., Dongwon, Y., Minji, A., Pere, M., Thomas, H., & Jinwoo, C., (2015). Tandem intercalation strategy for single-layer nanosheets as an effective alternative to conventional exfoliation processes. *Nature Communications, 6*(1), 1–7.
101. Lin, L., & Shaowei, Z., (2012). Creating high yield water soluble luminescent graphene quantum dots via exfoliating and disintegrating carbon nanotubes and graphite flakes. *Chemical Communications, 48*(82), 10177–10179.
102. Lin, L., Yaoxian, X., Shaowei, Z., Ian, M. R., Albert, C. M. O., & Dan, A. A., (2014). Fabrication and luminescence of monolayered boron nitride quantum dots. *Small, 10*(1), 60–65.
103. Hadz, S., Angerstein-Kozlowska, H., Vukovič, M., & Conway, B. E., (1978). Reversibility and growth behavior of surface oxide films at ruthenium electrodes. *Journal of The Electrochemical Society, 125*(9), 1471.
104. 104. Oh Si, H., & Linda, F. N., (2010). Direct synthesis of electroactive mesoporous hydrous crystalline RuO_2 templated by a cationic surfactant. *Journal of Materials Chemistry, 20*(19), 3834–3839.
105. Sassoye, C., Christel, L., Hung, L. K., Sophie, C., Cedric, B., Markus, A., & Clément, S., (2009). Block-copolymer-templated synthesis of electroactive RuO_2-based mesoporous thin films. *Advanced Functional Materials, 19*(12), 1922–1929.
106. Zheng, J. P., Cygan, P. J., & Jow, T. R., (1995). Hydrous ruthenium oxide as an electrode material for electrochemical capacitors. *Journal of the Electrochemical Society, 142*(8), 2699.
107. Augustyn, V., Jérémy, C., Michael, A. L., Jong, W. K., Pierre-Louis, T., Sarah, H. T., Héctor, D. A., et al., (2013). High-rate electrochemical energy storage through Li^+ intercalation pseudocapacitance. *Nature Materials, 12*(6), 518–522.
108. Raza, N., Tanuj, K., Vinamrita, S., & Ki-Hyun, K., (2021). Recent advances in bimetallic metal-organic framework as a potential candidate for supercapacitor electrode material. *Coordination Chemistry Reviews, 430*, 213660.

109. Liu, T., Liuyang, Z., Bei, C., Xianluo, H., & Jiaguo, Y., (2020). Holey graphene for electrochemical energy storage. *Cell Reports Physical Science, 1*(10), 100215.
110. Naguib, M., Murat, K., Volker, P., Jun, L., Junjie, N., Min, H., Lars, H., et al., (2011). Two-dimensional nanocrystals produced by exfoliation of Ti_3AlC_2. *Advanced Materials, 23*(37), 4248–4253.
111. Lukatskaya, M. R., Olha, M., Chang, E. R., Dall'Agnese, Y., Patrick, R., Pierre, L. T., Michael, N., et al., (2013). Cation intercalation and high volumetric capacitance of two-dimensional titanium carbide. *Science, 341*(6153), 1502–1505.
112. Kshetri, T., Duy, T. T., Huu, T. L., Dinh, C. N., Hien, V. H., Nam, H. K., & Joong, H. L., (2021). Recent advances in MXene-based nanocomposites for electrochemical energy storage applications. *Progress in Materials Science, 117*, 100733.
113. Huang, W., Lanping, H., Yanfeng, T., Zhongjian, X., & Han, Z., (2020). Recent advances in functional 2D Mxene-based nanostructures for next-generation devices. *Advanced Functional Materials, 30*(49), 2005223.
114. Yang, Q., Zhen, X., Bo, F., Tieqi, H., Shengying, C., Hao, C., Yingjun, L., et al., (2017). MXene/graphene hybrid fibers for high performance flexible supercapacitors. *Journal of Materials Chemistry A, 5*(42), 22113–22119.
115. Kumar, S., Malik, A. R., Sungwon, L., Minwook, K., Hyeryeon, H., Jun-Young, P., & Yongho, S., (2021). Supercapacitors based on Ti_3C_2Tx MXene extracted from supernatant and current collectors passivated by CVD-graphene. *Scientific Reports, 11*(1), 1–9.
116. Meng-Qiang, Z., Chang, E. R., Zheng, L., Maria, R. L., Chuanfang, Z., Katherine, L. V. A., Michel, W. B., & Yury, G., (2015). Flexible MXene/carbon nanotube composite paper with high volumetric capacitance. *Advanced Materials, 27*(2), 339–345.
117. Park, J. W., Dong, Y. L., Hyunsoo, K., Jae, S. H., Monica, J. De. A., Ray, H. B., & Seon, J. K., (2019). Highly loaded MXene/carbon nanotube yarn electrodes for improved asymmetric supercapacitor performance. *MRS Communications, 9*(1), 114–121.
118. Wang, R., Shaohong, L., Chen, X., Zhenyu, C., Houshen, L., Muhammad, A., Vincent, C., et al., (2021). MXene-carbon nanotubes layer-by-layer assembly based on-chip micro-supercapacitor with improved capacitive performance. *Electrochimica Acta, 386*, 138420.
119. Liang, W., & Igor, Z., (2021). MXene–carbon nanotube composite electrodes for high active mass asymmetric supercapacitors. *Journal of Materials Chemistry A, 9*(16), 10335–10344.
120. Li, L., Jianjian, F., Young-Rae, C., Je, M. Y., Yeon, S. J., Se, H. K., & Kwang Ho, K., (2021). Hierarchically layered nanocomposite electrodes formed by spray-injected MXene nanosheets for ultrahigh-performance flexible supercapacitors. *Applied Surface Science, 549*, 149226.
121. Rakhi, R. B., Bilal, A., Dalaver, A., & Husam, N. A., (2016). Direct chemical synthesis of MnO_2 nanowhiskers on transition-metal carbide surfaces for supercapacitor applications. *ACS Applied Materials & Interfaces, 8*(29), 18806–18814.
122. Hou, W., Yuan, S., Yu, Z., Tingting, W., Liqian, W., Youwei, D., & Wei, Z., (2021). Mixed-dimensional heterostructure of few-layer MXene based vertical aligned MoS_2 nanosheets for enhanced supercapacitor performance. *Journal of Alloys and Compounds, 859*, 157797.
123. Xie, W., Yanzi, W., Jie, Z., Meng, Z., Jiali, Y., Caizhen, Z., & Jian, X., (2020). MOF-derived $CoFe_2O_4$ nanorods anchored in MXene nanosheets for all pseudocapacitive flexible supercapacitors with superior energy storage. *Applied Surface Science, 534*, 147584.
124. Zhang, Y., Junming, C., Zeyu, Y., Lianjia, Z., Lili, W., & Wei, H., (2021). Assembling Co_3O_4 nanoparticles into MXene with enhanced electrochemical performance for advanced asymmetric supercapacitors. *Journal of Colloid and Interface Science, 599*, 109–118.

125. Gao, L., Chao, L., Weichun, H., Shan, M., Han, L., Qi, O., Ye, Z., et al., (2020). MXene/polymer membranes: Synthesis, properties, and emerging applications. *Chemistry of Materials, 32*(5), 1703–1747.
126. Zhan, X., Chen, S., Jian, Z., & Zhimei, S., (2020). MXene and MXene-based composites: Synthesis, properties, and environment-related applications. *Nanoscale Horizons, 5*(2), 235–258.
127. Cheng, W., Jimin, F., Haibo, H., & Derek, H., (2021). Interlayer structure engineering of MXene-based capacitor-type electrode for hybrid micro-supercapacitor toward battery-level energy density. *Advanced Science, 8*(16), 2100775.
128. Vahid, M. A., Jorge, M., Hengze, C., Emre, K., Jafar, O., Carlos, A. C., & Majid, B., (2018). Thick and freestanding MXene/PANI pseudocapacitive electrodes with ultrahigh specific capacitance. *Journal of Materials Chemistry A, 6*(44), 22123–22133.
129. Panda, S., Kalim, D., Khadheer, P. S. K., Jayaraman, T., Sivakumar, M., & Myong, Y. C., (2022). MXene based emerging materials for supercapacitor applications: Recent advances, challenges, and future perspectives. *Coordination Chemistry Reviews, 462*, 214518.
130. Xie, Y., Haitao, Z., Haichao, H., Zixing, W., Zhong, X., Haibo, Z., Yuchen, W., et al., (2020). High-voltage asymmetric MXene-based on-chip micro-supercapacitors. *Nano Energy, 74*, 104928.
131. Wang, Y., Xue, W., Xifei, L., Xiaolong, L., Yang, L., Yang, B., Huanhao, X., & Guohui, Y., (2021). A high-performance, tailorable, wearable, and foldable solid-state supercapacitor enabled by arranging pseudocapacitive groups and MXene flakes on textile electrode surface. *Advanced Functional Materials, 31*(7), 2008185.
132. Chen, W., Daotong, Z., Kai, Y., Min, L., Pei, Y., & Xiaoyan, Z., (2021). Mxene (Ti_3C_2Tx)/cellulose nanofiber/porous carbon film as free-standing electrode for ultrathin and flexible supercapacitors. *Chemical Engineering Journal, 413*, 127524.
133. Zheng, J., Xin, P., Xiaomin, H., Dongbin, X., Yang, S., Xiaoxia, L., Ning, W., et al., (2020). Integrated $NiCo_2$-LDHs@ MXene/rGO aerogel: Componential and structural engineering towards enhanced performance stability of hybrid supercapacitor. *Chemical Engineering Journal, 396*, 125197.
134. Li, L., Weijia, L., Kai, J., Di, C., Fengyu, Q., & Guozhen, S., (2021). In-situ annealed Ti_3C_2Tx MXene based all-solid-state flexible Zn-ion hybrid micro supercapacitor array with enhanced stability. *Nano-Micro Letters, 13*(1), 1–11.
135. Zhang, M., Jun, C., Yi, W., Jia, S., Tianci, J., Yanyu, Z., Weimeng, S., et al., (2021). Electrolyte-mediated dense integration of graphene-MXene films for high volumetric capacitance flexible supercapacitors. *Nano Research, 14*(3), 699–706.
136. Li, X., Yanan, M., Yang, Y., Guosheng, L., Chuankun, Z., Minglei, C., Yongchen, X., et al., (2022). A flexible Zn-ion hybrid micro-supercapacitor based on MXene anode and V_2O_5 cathode with high capacitance. *Chemical Engineering Journal, 428*, 130965.
137. Zhu, Ch., & Fengxia, G., (2021). Macroscopic MXene ribbon with oriented sheet stacking for high-performance flexible supercapacitors. *Carbon Energy, 3*(1), 142–152.
138. Sikdar, A., Pronoy, D., Sujit, K. D., Abhisek, M., Padma, N., Subhradip, G., & Uday, N. M., (2021). Spontaneous three-dimensional self-assembly of MXene and graphene for impressive energy and rate performance pseudocapacitors. *Electrochimica Acta, 391*, 138959.
139. Liu, Y. M., Céline, M., & Berend, S., (2019). Carbons with regular pore geometry yield fundamental insights into supercapacitor charge storage. *ACS Central Science, 5*(11), 1813–1823.

140. Miranda, D. A., & Paulo, R. B., (2016). Density functional theory and an experimentally-designed energy functional of electron density. *Physical Chemistry Chemical Physics, 18*(37), 25984–25992.
141. Mišković, Z. L., & Nitin, U., (2010). Modeling electrolytically top-gated graphene. *Nanoscale Research Letters, 5*(3), 505–511.
142. Ilani, S., Luke, A. K. D., Mark, K., & McEuen, P. L., (2006). Measurement of the quantum capacitance of interacting electrons in carbon nanotubes. *Nature Physics, 2*(10), 687–691.
143. Bisquert, J., (2003). Chemical capacitance of nanostructured semiconductors: Its origin and significance for nanocomposite solar cells. *Physical Chemistry Chemical Physics, 5*(24), 5360–5364.
144. Bueno, P. R., (2019). Nanoscale origins of super-capacitance phenomena. *Journal of Power Sources, 414*, 420–434.
145. Yang, G. M., Zhang, H. Z., Fan, X. F., & Zheng, W. T., (2015). Density functional theory calculations for the quantum capacitance performance of graphene-based electrode material. *The Journal of Physical Chemistry C, 119*(12), 6464–6470.
146. Sharma, A., Abhinandan, P., Namsheer, K., Pratap, M., Brahmananda, C., & Chandra, S. R., (2021). All-solid-state asymmetric supercapacitors based on VS4 nano-bundles and MXene nanosheets. *Journal of Materials Science, 56*(36), 20008–20025.
147. Raj, K. A. S, Pratap, M., Sithara, R., Brahmananda, C., & Chandra, S. R., (2022). Heterostructured metallic 1T-VSe_2/Ti_3C_2T x MXene nanosheets for energy storage. *ACS Applied Nano Materials,5*(3), 4423–4436.
148. Zhang, J., Chengfeng, D., Zhengfei, D., Wei, C., Yun, Z., Bing, L., Yun, Z., et al., (2017). NbS_2 nanosheets with M/Se (M= Fe, Co, Ni) codopants for Li^+ and Na^+ storage. *ACS Nano, 11*(10), 10599–10607.
149. Zhang, Q., Wei, W., Yujiao, W., Pingyuan, F., Kangli, W., Shijie, C., & Kai, J., (2016). Controllable construction of 3D-skeleton-carbon coated $Na_3V_2(PO_4)_3$ for high-performance sodium ion battery cathode. *Nano Energy, 20*, 11–19.
150. Jayaseelan, S. S., Sivaprakasam, R., Balasubramaniam, S., Min-Kang, S., Myung-Seob, K., Hak-Yong, K., & Byoung-Suhk, K., (2018). Mesoporous 3D $NiCo_2O_4$/MWCNT nanocomposite aerogels prepared by a supercritical CO_2 drying method for high performance hybrid supercapacitor electrodes. *Colloids and Surfaces A: Physicochemical and Engineering Aspects, 538*, 451–459.
151. Pathak, M., Jeena, R. J., Brahmananda, C., & Chandra, S. R., (2020). High performance supercapacitor electrodes based on spinel $NiCo_2O_4$@ MWCNT composite with insights from density functional theory simulations. *The Journal of Chemical Physics, 152*(6), 64706.

CHAPTER 2

Two-Dimensional (2D) Materials-Based Nanostructured Hybrid Composites: A Newer Paradigm for Energy Application

NIDHI SHARMA,[1] NEETU TALREJA,[2] DIVYA CHAUHAN,[3] and MOHAMMAD ASHFAQ[4]

[1]*Department of Physics, Dhanauri P.G. College, Dhanauri, Roorkee Haridwar, Uttarakhand, India*

[2]*Department of Science, Faculty of Science and Technology, Alliance University, Anekal, Bangalore, Karnataka, India*

[3]*Department of Chemical and Biomedical Engineering, University of South Florida, Tampa, Florida, United States*

[4]*The Department of Biotechnology and University Center for Research and Development (UCRD) is located at Chandigarh University, Gharaun, Mohali, Punjab, India*

ABSTRACT

Presently, two-dimensional nanomaterials (2D-NM) are continuously gaining attention from researchers due to their extraordinary properties, such as high mechanical strength, high surface area, and electrical and optical properties. These properties make them promising candidates for various applications. The high surface area, atomic thickness, and tunable ability of 2D-NM offer advantages in terms of their applicability. The mechanical, electrical, and thermal properties of 2D-NM, along with their unique characteristics, greatly influence their potential applications and overall applicability. This mini-review article focuses on the synthesis processes, composition, and

Sustainability in Energy and Environment: Engineered Materials and Smart Computational Techniques. Ann Rose Abraham, Heru Susanto, A. K. Haghi, & Kian Hariri Aslı (Eds.)

crystal structure of 2D-NM. Additionally, the role of 2D-NM-based hybrid composites in solar cells and supercapacitor applications, based on their characteristics, is discussed. In conclusion, future challenges in the field of 2D-NM-based hybrid composites are addressed, taking into account their characteristic properties.

2.1 INTRODUCTION

Nanomaterials (NM) have emerged as next-generation materials that encompass a wide range of sizes (1–100 nm) and dimensions [zero (0D), one (1D), two (2D), and three (3D)]. The unique characteristics of NM, particularly their high surface area, mechanical, electrical, optical, catalytic, and thermal properties, make them promising and suitable candidates for numerous applications such as environmental treatments, energy, electrical, solar cells, agriculture, medicine, drug delivery, and catalysts. Furthermore, the properties of NM can be easily modified or adjusted by changing their surface functional groups, synthesis parameters, size, shape, and surface charges [1–5]. NM are mainly classified into four types based on dimensionality: (i) 0D-NM (particles and clusters); (ii) 1D-NM (nanotubes, nanofibers, and nanowires); (iii) 2D-NM (plates and sheets); and (iv) 3D-NM (bulk powder and bundles of nanotubes/wires). Recently, 2D-NMs have gained significant research interest since the discovery of graphene [6, 7]. Graphene is the first well-known 2D-NM with a zero-band gap value attributed to high electron mobility and exceptional conductivity. The extraordinary characteristics such as mechanical, optical, electronic, and thermal transport make them appropriate candidates for energy and electronics. Moreover, the tunable atomic thickness and high surface area of 2D-NMs make them promising candidates for end applications [8–13]. In one aspect, academicians/scientists continue to concentrate on newer 2D-NMs to improve their efficacy towards end applications.

2D-NMs, such as transition metal dichalcogenides (TMDCs) (WS_2, WSe_2, MoS_2, and $MoSe_2$), graphene, hexagonal boron nitride (h-BN), metal oxides, transition metal halides (TMHs), graphitic carbon nitride (g-C_3N_4), MXene, layered-double hydroxides (LDHs), and black phosphorus (BP), have been synthesized for numerous applications such as energy, solar cells, sensors, antibacterial agents, environmental remediation, photocatalysis, agriculture, electronics, supercapacitors (SCs), biomedicine, and wearable devices [14–21]. Importantly, the thermal, electrical, and mechanical properties of 2D-NMs directly influence their potential applications. Therefore, understanding the properties of 2D-NMs, particularly their thermal, electrical, and mechanical properties, can be advantageous in enhancing their efficacy. However, the thickness of the 2D-NM is too thin to absorb adequate light, which limits its effectiveness mainly in photocatalysis, optoelectronics, and optical sensors. Approximately 2.3% and 5.6% light absorption were observed in graphene, and the MoS_2 layer was significantly lower than the bulk materials. Therefore, there is a necessity to improve the optical and optoelectronics properties to meet the requirement [22–25]. The development of 2D-NM offers new paradigms due to its exceptional potential ability in most areas of research.

Researchers continue focusing on developing the 2D-NM-based hybrid system to improve the electrical, mechanical, and thermal characteristics by incorporating metal, non-metals, and polymers within the 2D-NM. Containing the metal nanoparticles (M-NPs) within the 2D-NM like graphene improved the photon absorption, conductivity, mechanical stability, and thermal properties, thereby improving the applicability for end applications [7, 26, 27]. Moreover, 2D-NM-based hybrid materials might be tuned according to the applications, thereby easily improving the materials' efficacy via changing the properties of 2D-NM. The materials science, engineering, chemistry, and nanotechnology offer newer paradigms for developing a newer class of materials, namely 2D-NM that contains numerous properties, mainly electrical, thermal, and mechanical properties, which are required for numerous end applications [28–31]. However, there is a knowledge gap between the characteristics and properties of 2D-NM and their chemistry. With advancements in material science and chemistry, researchers are attempting to develop newer 2D-NM materials for various applications, particularly in environmental remediation, sensors, energy systems, and drug delivery systems. Furthermore, exploring the role of electrical, thermal, and mechanical properties in their application is still desirable for better understanding and utilization [32–35]. This chapter focuses on synthesizing 2D-NM, primarily discussing their electrical, thermal, and mechanical properties. Additionally, we will explore the application of 2D-NM in solar cells and supercapacitors, considering their characteristics.

2.2 SYNTHESIS OF 2D-NM

The synthesis of 2D-NMs is challenging due to their isotropic crystal nature and high surface energy. These materials and their derivatives are primarily synthesized using two main methods: (i) the top-down process, where bulk materials are converted into smaller or nano-sized particles; and (ii) the bottom-up process, where atoms/molecules are assembled into nano-sized materials. There are two main processes used for the synthesis of 2D-NMs: (a) exfoliation process; and (b) chemical vapor deposition (CVD) process.

2.2.1 EXFOLIATION PROCESS

Graphene and other 2D-NM were exfoliated using a liquid-phase exfoliation process. The liquid-phase exfoliation process produced graphene from graphite using NMP solvent. Usually, bulk graphite is detached into single/multilayer graphene due to the generated force. The degree of graphene exfoliation is directly related to the dispersion grade and stability. The graphene layers are attached using Vander-Waal force and functional groups between two adjacent layers. The exfoliation process is facile and scalable with a high grade of 2D-NM. The high quality and yields of 2D-NM mainly depend on the organic solvents, sonication times, concentration of the materials, integration of surfactant, and polymers [36–39]. Several 2D-NM have been synthesized using the exfoliation process. For instance, Guan et al. [40] synthesized MoS_2 nanosheets using a liquid-phase exfoliation process. The sodium dodecyl benzene sulfonate surfactant in water was used as a solvent for synthesizing MoS_2 nanosheets. The synthesized MoS_2 nanosheets were highly stable, and few-layer MoS_2 nanosheets were controlled with the surfactant concentration [40]. Tan et al. synthesized MoS_2 nanosheets using a liquid-phase exfoliation process. The supercritical CO_2-modified N-methyl-2-pyrrolidone (NMP) solvent was used as a solvent. The produced MoS_2 nanosheets are highly stable and consist of single/few-layer nanosheets [41]. Sahoo et al. synthesized MoS_2 nanosheets for photocatalytic applications. The MoS_2 nanosheets were synthesized through an exfoliation process, using acetone as a solvent [42]. Ashfaq et al. synthesized 2D-Cu-WS_2 nanosheets using a simple exfoliation process for biomedical applications. WS_2 nanosheets were synthesized through an exfoliation process in NMP solvents [14]. Furthermore, the exfoliation process easily separates the adjacent graphene layers, allowing for the production of single/multilayer graphene. Additionally, we can easily control the shape, size, crystal structure, and stability of the 2D-NM using solvents, surfactants, and material concentrations. Figure 2.1 illustrates the exfoliation process of the 2D-NM.

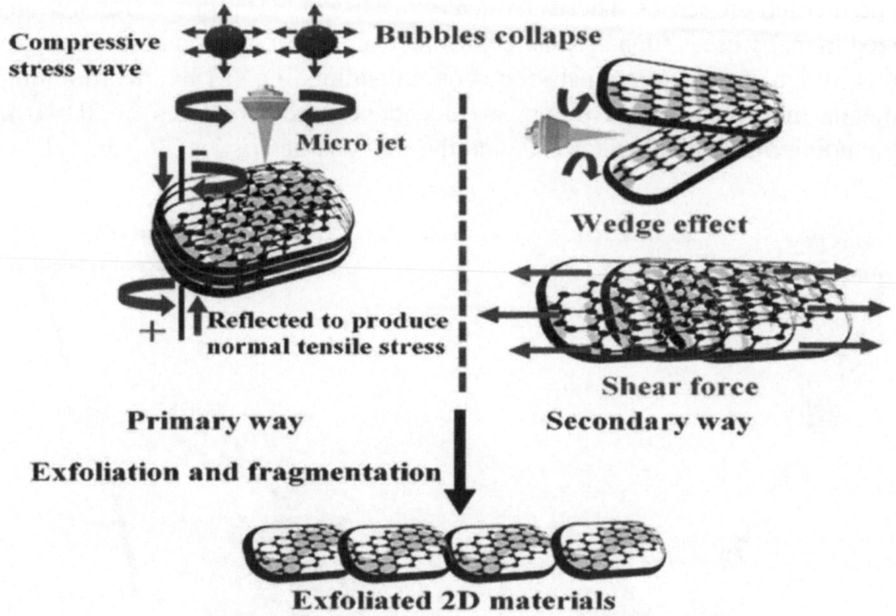

FIGURE 2.1 A schematic illustration of the exfoliation process of 2D-NM.
Source: The image was incorporated with permission [37].

2.2.2 CVD PROCESS

Usually, the CVD process synthesizes numerous nanomaterials (NMs) such as graphene, carbon nanofibers (CNFs), carbon nanotubes (CNTs), and their derivatives. Cu or Ni foil or silicon wafers are used as a substrate for growing NMs, including graphene. The substrate is placed within the CVD reactor, and the precursor is exposed to the substrate surface at high temperatures using inert conditions. The precursor decomposes on the substrate to synthesize 2D-NM [37, 43, 44]. Moreover, the layers or thickness of the 2D-NM can be controlled by tuning the reaction conditions. Figure 2.2 shows the schematic representation of the CVD process.

2.3 PROPERTIES OF 2D-NM

The 2D-NM-like graphene is wide open, with prodigious potential in numerous end applications due to its extraordinary electrical, mechanical, and thermal properties. These characteristics of 2D-NM might be revolutionized in many areas such as solar cells, electronics, sensors, and SCs. Despite this, researchers continuously focus on amending 2D-NM by incorporating metals, non-metals, and polymers to enhance the properties of 2D-NM, including mechanical, electrical, and thermal characteristics [26, 46, 47].

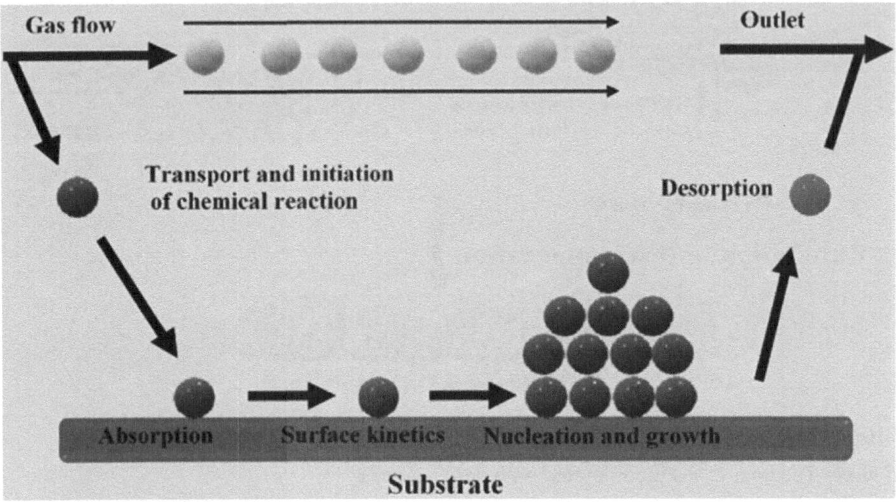

FIGURE 2.2 A schematic representation of the CVD process.

Source: The image was incorporated with permission [37].

2.3.1 THERMAL PROPERTIES

2D-NM, including graphene, is continuously capturing researchers' interest due to its thermal properties, particularly thermal conductivity. The thermal properties of 2D-NM depend on various factors such as size, isotopic doping, surface functional group, substrate, and edge configuration. These factors can be modified to easily tune the thermal properties. Understanding these thermal properties can be beneficial and directly related to the application of 2D-NM. Additionally, the electrical conductivity of NM is similar to that of their bulk materials. However, the thermal conductivity of NM is significantly reduced compared to that of bulk materials due to photon scattering. Based on this concept, 2D-NM shows promise as a candidate for determining thermal conductivity performance. Research has been conducted so far on modifying the thermal properties of 2D-NM. For example, Judek et al. focus on theoretical analysis to determine thermal conductivity. The data suggest that the laser beam can distribute power in all directions and provide new strategies for the theoretical and experimental data acquisition of the Opto-thermal method. Moreover, the Opto-thermal method is applicable for 2D-NM, which easily characterizes various thermal properties [48]. Feng et al. focus on metal-doped phosphorene to augment thermal stability. The data suggest that Al doping within the phosphorene increases thermal properties by two times. Moreover, Al doping increases the structural ability of 2D-NM, thereby enhancing thermal stability. The high thermal and structural stability of 2D-NM is of great interest in nanoelectronics applications [49]. Tang et al. synthesize vermiculite sheets (VS) and incorporate them in PEO-LiTFSI to produce a VS-PEO-LiTFSI composite. The data indicate that incorporating PEO-LiTFSI within the VS-PEO-LiTFSI composite considerably increases the thermal stability of the composite [50]. Krishnamoorthy et al. use non-equilibrium molecular dynamics to calculate the thermal stability of $MoSe_2$ and WSe_2 by designing lattices. The data suggest that doping with metal designs heterostructures that ultimately increase thermal stability. Moreover, the thermal stability of 2D-NM might be controlled by changing the concentration of the dopant [51]. Tong et al. report the synthesis of Beryllonitrene and N-doped Beryllonitrene to determine thermal conductivity. The data suggest that doping increases the carrier and electrical density, thereby increasing the electrical and thermal properties of Beryllonitrene [52]. 2D-NM are mostly semiconductor materials in which heat is transported by photons. Heat conduction is very efficient because of the carbon atoms that contain the sp^2 bond. The high thermal conductivity of graphene can be enhanced by introducing a small gap at its center. In order to understand the exact mechanism of the thermal properties of 2D-NM and improve its applicability and device performance, it is necessary to investigate further. This knowledge will enable us to design different 2D-NM and their related devices with improved thermal management [53–56]. The thermal properties primarily depend on the atomic structures, which can be adjusted by incorporating/changing the surface functional group, applying strains, doping with metals/non-metals, ensuring structure stability, and optimizing lattice design. Existing literature suggests that modifying the lattice and doping with metals/non-metals can enhance the structural stability and, consequently, the thermal stability of 2D-NM. By manipulating the lattice design and conducting metal/non-metal doping, we can easily modify the performance of 2D-NM in various applications.

2.3.2 ELECTRICAL PROPERTIES

The electrical properties of 2D-NM are mostly determined using an electronic band structure, which establishes the energy states in relation to the crystalline structure. 2D-NM-based devices, such as solar cells and gas sensors, exhibit their own electronic characteristics. Numerous studies have been conducted to enhance the electrical properties of 2D-NM. For instance, Sreeprasad et al. synthesized nanosheets of Au-incorporated MoS_2 (Au-MoS_2) using chemical and microwave processes, and investigated their electrical, thermal, and structural properties. The data indicate that the incorporation of Au within MoS_2 significantly improves its electrical, thermal, and structural properties. Figure 2.3 illustrates the conductivity of MoS_2 and Au-MoS_2. The conductivity of Au-MoS_2 increases with temperature. In fact, the conductivity of Au-MoS_2 is approximately 10^3 times higher than that of MoS_2[57].

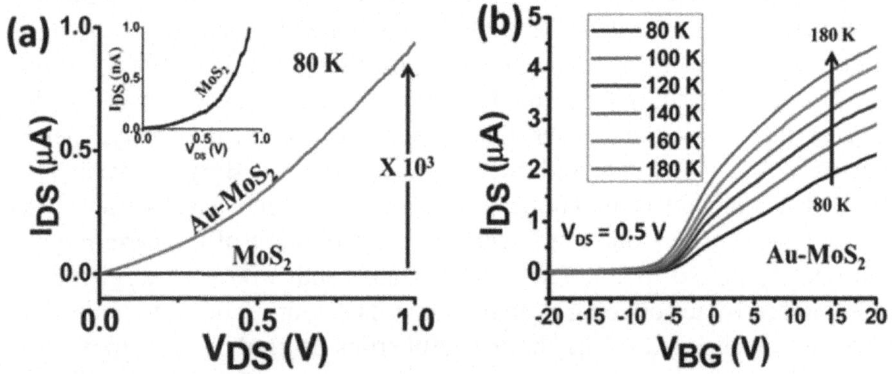

FIGURE 2.3 Conductivity of MoS_2 and Au-MoS_2 (a) at 80 K; and (b) at 80–180 K.

Source: The image was taken with permission [57].

Yang et al. focused on metal doping within Sc_2CT_2 to enhance the magnetic and electronic properties of 2D-NM based on Sc_2CT_2. The data suggested that transition metals (TMs) such as Ti, Cr, Mn, and V could be doped into Sc_2CT_2-based 2D-NM. The doping of TMs improved the magnetic and electronic characteristics of Sc_2CT_2-based 2D-NM [58]. Xiong et al. theoretically examined the role of surface functional groups on Sc_2CT_2 MXene-based 2D-NM. The data suggested that surface functionalization of Sc_2CT_2 MXene-based 2D-NM exhibited high redox ability of photoactive carriers, high absorption, carrier mobility, and thus high photocatalysis [59]. Vargas-Bernal focused on the 2D-NM-based gas sensor using electrical properties. The literature suggested that the 2D-NM-based gas sensor primarily exhibited hexagonal crystalline structure, while some materials also showed monoclinic, orthorhombic, and triclinic structures. The sensing response was mainly based on the electrical properties of 2D-NM [60]. Figure 2.4 illustrates the schematic design of the gas sensor based on electrical properties. Postorino et al. investigated the effect of biaxial uniform strain on the electrical and structural properties of MoS_2 and $MoTe_2$-based 2D-NM using first principle DFT calculations, with a focus on ionization potential, electron affinity, work function, and band structure. The data suggested that biaxial uniform strain effectively tunes the structural and electrical properties. Moreover, the work function always increases in monolayer 2D-NM, whereas the ionization potential and electron affinity exhibit material-dependent behavior [61]. Cui et al. focused on tuning the electronic properties of p and n-type doped $MoSi_2N_4$-based 2D-NM obtained by first principle calculations. The data suggested that molecular doping (p and n-type doping) modifies the electronic properties. The tuning of the electronic properties of p and n-type doped $MoSi_2N_4$-based 2D-NM could enhance their applicability [62]. The aforementioned data indicate that 2D-NM have inherent electrical properties. Moreover, the electrical properties can be easily tuned by structural changes of the 2D-NM. The structural changes achieved through the use of metals, non-metals, and surface functionalization offer advantages in numerous applications, including solar cells and sensors.

2.3.3 MECHANICAL PROPERTIES

The mechanical properties of the 2D-NM provide exciting opportunities for the development and design of a new class of 2D-NM with high flexibility and in-plane stiffness. The high flexibility and in-plane stiffness of the 2D-NM make them a promising candidate for various applications, particularly wearable devices. It is important to note that the mechanical properties of the 2D-NM and their derivatives do not directly impact the performance of the respective applications, but they are necessary for real-time applications like wearable electronic devices that require high stiffness and mechanical strength [56, 63, 64].

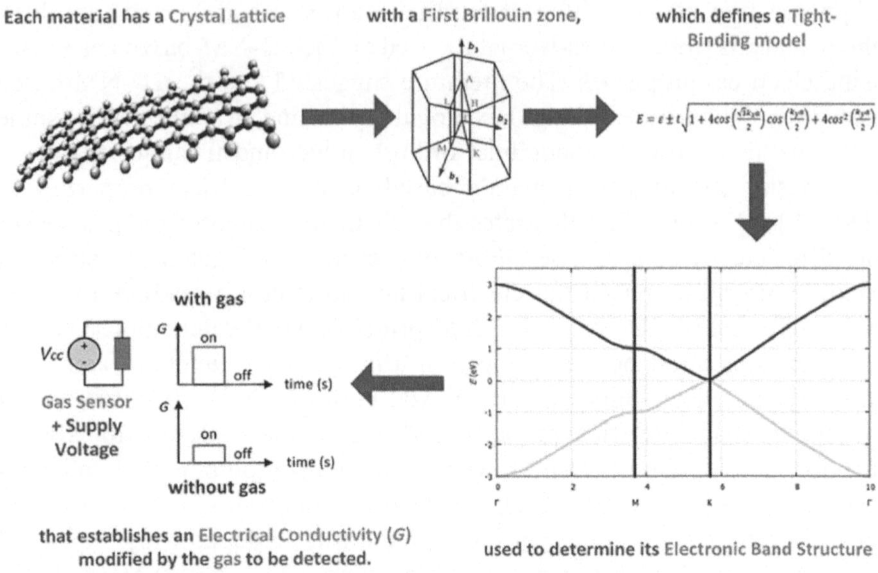

FIGURE 2.4 A schematic illustration of the gas sensor design based on electrical properties.
Source: The image was taken with permission [60].

2.4 2D MATERIALS FOR SOLAR CELL APPLICATION

2D-NM-based materials are attracting interest in solar cell applications due to their superior thermal, mechanical, electronic, and optical properties. Currently, nano-dimensional thin layers are in high demand as they offer flexibility, high performance, and mechanical strength. Additionally, these materials are a logical choice for advanced and cost-effective solar cell flexible devices due to their efficient utilization of solar energy. Various 2D-NM, such as graphene, graphene oxide, TMDs, boron, carbon nitride, and metallic nanosheets, are being used in various applications, including solar cells, because of their high conductivity, transparency, and environmental stability. Different functional solar cells, such as perovskites, organic solar cells, and dye-sensitized solar cells, have utilized 2D-NM as electron transport layers, hole transport layers, or absorber layers to achieve high performance. However, commercialization still poses a major challenge due to concerns regarding the stability of 2D-NM. Various researchers have reported several methods to achieve stability and increase the lifespan of 2D-NM/nanosheets. For example, Krishnamoorthy et al. synthesized graphene-incorporated WS_2 nanosheets using a hydrothermal process. The coated composite demonstrates an excellent power conversion efficiency (PCE) of 9.6%, with a fill factor of 0.66 [65]. Zhang et al. synthesized Cu_2ZnSnS_4-based nanosheets with high energy (002) facets. These exposed facets demonstrate good photo electrochemical performance with a PCE of 6.68%. The design of an eco-friendly Cu-based wurtzite structure is a promising candidate for solar cell applications [66]. Figure 2.5 shows the schematic illustration of the synthesized Cu_2ZnSnS_4-based nanosheets, XRD, and Raman spectra.

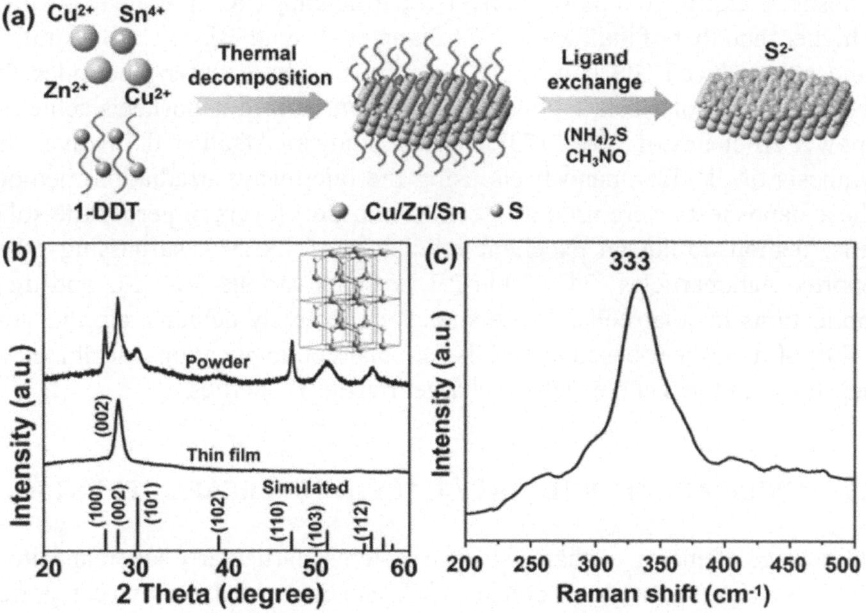

FIGURE 2.5 (a) A schematic illustration of synthesized Cu_2ZnSnS_4-based nanosheets; (b) XRD spectra; and (c) Raman spectra.

Source: The image was incorporated with permission [66]; Creative Commons license.

Fang et al. synthesized a CoS nanosheet using a hydrothermal process. The synthesized nanosheet was dispersed with PEDOTPSS to form a CoS/PEDOT PSS suspension. The fabricated device demonstrates a power efficiency of 11.2% [67]. Mahesh et al. synthesized 2D-ZnO nanosheets (2D-ZnO-NS) incorporated with beta-substituted porphyrins photosensitizer. The 2D-ZnO-NS was produced using the hydrothermal process. The synthesized nanohybrid shows a power efficiency of 0.16%, higher than the 2D-ZnO-NS [68]. Another study by Jiang et al. synthesized S-doped Co_3O_4/graphene using a wet chemical process. The synthesized material shows a power efficiency of 8.24% using visible light irradiation [69]. Huang et al. synthesized Ultrathin $FeSe_2$ Nanosheets using the phosphine-free hot-injection method. The power efficiency was 7.53%, comparable to the Pt-based device [70]. Sheng et al. synthesized 2D-TiO_2-nanosheet using the self-assembly method for the photovoltaic (PV) device. The data suggested that the 2D-TiO_2-nanosheet-based dye-sensitized solar cells have a power efficiency of 8.28% [71]. Dharamalingam et al. synthesized MoS_2 nanosheets and applied them in dye-sensitized solar cell applications. The nanosheets exhibited a power conversion efficiency (PCE) of 3.81%, which is higher than that of bulk MoS_2 [72]. Jiang et al. synthesized Co_3S_4 ultrathin nanosheet/reduced-GO (Co_3S_4-rGO) composites as counter electrodes for dye-sensitized solar cell applications. The synthesized nanosheets achieved a power efficiency of 8.08% [73]. Another study by Afzali et al. involved the synthesis of 2H-TaS_2 nanosheets using the microwave irradiation method. These nanosheets were used as electron transport layers in perovskite solar cells, demonstrating an excellent efficiency of 15.23%, surpassing other reported nanoparticles [74]. Table 2.1 presents various 2D-NMs and their applications in solar cells. These studies collectively demonstrate the suitability of nanosheet-based materials for solar cell applications, attributed to their low band gap and excellent charge transfer properties.

2.5 APPLICATION OF 2D MATERIALS IN SUPERCAPACITORS (SCS)

The storage of energy in charge storage systems, particularly supercapacitors (SCs), is a significant global concern. SCs can be divided into two categories –(i) electrochemical double layer capacitors (EDLCs), and (ii) pseudo capacitors. EDLCs utilize an electrode-electrolyte double layer to store energy, while pseudo capacitors employ redox reactions on the electrode surface for charge storage. Various materials have been used for supercapacitor applications, including carbon-based materials, metal oxides/hydroxides, and polymer metal/carbon composites. Prominent carbon-based materials used as active electrode materials in EDLCs include activated carbon (AC), mesoporous carbon, carbon nanofibers (CNFs), carbon nanotubes (CNTs), and graphene. Additionally, several metal oxides and conducting polymer (CP) composites are used for pseudo capacitors. Carbon-based materials mainly work based on the EDLC mechanism, as pore size distribution is an important factor for EDLC. Therefore, they often exhibit very high power output and better cycling ability. However, EDLCs have lower energy density values than pseudo capacitors, since pseudo capacitors involve active redox materials to

TABLE 2.1 Different 2D-NM and Their Solar Cell Application

SL. No.	2D Nanosheet	Synthesis Method	Performance	References
1.	2D-ZnO nanosheets-porphyrins.	Hydrothermal	PCE 0.13%	[68]
2.	CoS nanosheets-Si/ PEDOT:PSS	Hydrothermal	PCE 11.2%	[67]
3.	$FeSe_2$nanosheets	Phosphine-free hot-injection method.	PCE 7.53%	[70]
4.	2D-TiO_2 nanosheet	Molecular self-assembly using the micelle-assisted solution method.	PCE 8.28%	[71]
5.	Cu_2ZnSnS_4 nanosheets	Thermal decomposition.	PCE 6.68%,	[66]
6.	S-Co_3O_4/r-GO	Wet chemical	PCE 8.24%	[69]
7.	Co_3S_4-r-GO	Non-injection hydrothermal process.	PCE 8.08%	[73]
8.	2H-TaS_2 nanosheets	Microwave irradiation method.	PCE 15.23%	[74]
9.	Graphene-WS_2	Hydrothermal	Open circuit voltage (Jsc) of 0.79 mV, short circuit current (Voc) of 18.6 mA cm^{-2}, fill factor – 0.66, PCE 9.6%.	[65]
10.	MoS_2nanosheets	Sodium dodecyl sulfate mediated hydrothermal.	PCE 3.81%	[72]

store charge both on the surface and in the sub-surface layer. Recently, 2D materials like graphene, TMDs, and MXene have demonstrated excellent charge storage performance due to their high electrical conductivity, surface-to-charge ratio, and flexible functionality. These materials are attracting interest in energy storage applications. Additionally, high performance and portability are major requirements for next-generation wearable devices.

Several researchers are reporting the application of 2D materials in advance of supercapacitor (SC) application. For instance, Huang et al. synthesized composite materials (P-G-MoS_2) based on polyaniline (PANI), graphene, and MoS_2 using a polymerization process for supercapacitor application. The prepared P-G-MoS_2 is effectively used in supercapacitor applications with a specific capacitance of 575 F/g at A/g. Moreover, the high specific capacitance is mainly due to the synergetic effects of the polymer, graphene, and MoS_2 [75]. Quan et al. synthesized a composite of α-Fe_2O_3 nanoplates and reduced graphene oxide (Fe-rGO) for supercapacitor application. The material was synthesized using a one-pot hydrothermal synthesis method, which further applied to SCs application and showed an excellent specific capacitance of 903 F/g at a current density of 1 A/g [76]. Ansari et al. mechanically exfoliated MoS_2 nanosheets and incorporated them with conductive polyaniline (P-MoS_2) for supercapacitor application. The data indicates that the produced P-MoS_2-based composite materials are effectively used for supercapacitor applications, with a specific capacitance of 510 F/g. The extraordinary specific capacitance is mainly due to the synergetic effects of conductive polymers and 2D-MoS_2 nanosheets [77]. Navarro-Suárez et al. synthesized titanium carbide and rGO-based 2D-NM composite nanolayers for supercapacitor applications. The synthesized composite shows a specific capacitance of 254 F g^{-1} at a scan rate of 2 mV s^{-1}. However, with further increase in the scan rate up to 100 mV s^{-1}, the specific capacitance decreases to 193 F g^{-1} [78]. Rajendran et al. synthesized VS_4 nanosheets using a hydrothermal method, which demonstrates a specific capacitance of 215 F g^{-1} at a scan rate of 5 mV s^{-1} with excellent retention of 82% within 2,000 cycles [79]. Chang et al. synthesized $NiSe_2$ nanosheets with selenium vacancy, which showed excellent performance (328 Fg^{-1}) towards charge storage [80]. Using the nitridation method, Yuan et al. synthesized 2D carbon-wrapped TiN (2D-C-TiN). The wrapped nanolayer exhibited excellent performance of 238 F/g at a current density of 0.5 A/g and a high-rate capacitance [81]. Geng et al. synthesized B/N co-doped 2D porous carbon nanosheets (B/N-C-NS) by simply mixing and carbonizing a mixture of ammonium humate in H_3BO_3, which demonstrated excellent specific capacitance at a current density of 0.2 A/g. Furthermore, upon increasing the current density up to 100 A/g, the material exhibited a capacitance of 125 F/g with a cycling retention of 94.8% after 10,000 cycles [82]. In another study by Zhang et al., an Nb_2SnC nanolayer was used for supercapacitor applications. The synthesized nanolayers showed a capacitance of 140 F/g with an energy density of 14.4 Whg^{-1} at a power density of 36 W kg^{-1}, along with high cycling stability [83]. Naz et al. synthesized 1T-MoS_2 nanosheets on 3D-r-GO (MoS_2-r-GO) using a one-pot hydrothermal process. The synthesized material exhibited excellent performance of 442 F/g at a current density of 1 A/g, with exceptional cycle stability and retention [84]. Venkateshalu et al. synthesized 2D-vanadium-based MXene and applied it for supercapacitor applications. The material was synthesized using high-temperature annealing, etching of the MXene layer, and slurry deposition on the CF surface. The material exhibited a capacitance of 112.8 F/g at a current density of 1.85 mA/cm^2 using an aqueous electrolyte [85]. Figure 2.6 illustrates the schematic illustration of the 2D-vanadium-based MXene and its supercapacitor application.

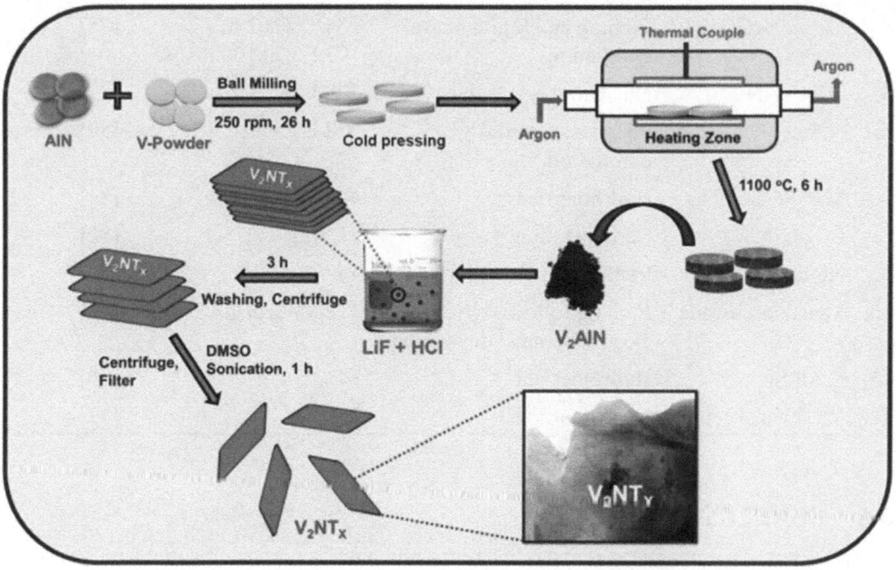

FIGURE 2.6 A schematic illustration of the 2D-vanadium-based MXene and its supercapacitor application.

Source: The image was taken with permission [85].

Alam et al. synthesized a $Co(OH)_2$–$MoSe_2$-based hybrid (C-$MoSe_2$) nanocomposite using a single-step hydrothermal reaction and applied it for asymmetrical supercapacitor applications. The synthesized material exhibits a specific capacitance of 541.55 F/g at a current density of 1 A/g. However, with a further increase in current density to 10 A/g, the specific capacitance decreases to 397 F/g, while still maintaining an excellent retention capacity of 86.2%. The material also demonstrates an energy density of 18.28 Wh/kg at a maximum power density of 12.209 kW kg^{-1} [45]. Table 2.2 summarizes different 2D-NM and their applications in SCs. These studies collectively highlight the excellent charge storage capabilities of nanosheets, although their stability remains a concern.

TABLE 2.2 Different 2D-NM and Their Supercapacitor Applications

SL. No.	2D-NM	Synthesis	Capacitance Value	References
1.	P-G-MoS_2	Polymerization	575 F/g at A/g	[75]
2.	Fe-rGO	Hydrothermal	903 F/gat 1 A/g	[76]
3.	P-MoS_2	Mechanical exfoliation	510 F/g at A/g	[77]
4.	Carbide-rGO	Etching and liquid phase exfoliation	254 F/g at 2 mV/s 193 F/g at 100 mV/s	[78]
5.	2D-C-TiN	Nitridation	238 F/g	[81]
6.	$NiSe_2$	Plasma-assisted dry exfoliation	328 F/g	[80]
7.	MoS_2-r-GO	Hydrothermal	442 F/g at 1 A/g	[84]
8.	B/N-C-NS	Carbonization	311 F/g	[82]
9.	Nb_2SnC	Etching/delamination	140 F/g	[83]
10.	Vanadium nitride	Etching followed by electrochemical deposition	112.8 F/g at 1.85 mV/s	[85]
11.	C-$MoSe_2$	Hydrothermal	541.55 F/g	[45]
12.	VS_4 nanosheets	Hydrothermal	215 F/g	[79]

2.6 CONCLUSION

Advancements in characterization tools and increasing demands for innovative technologies have led to the development of newer 2D-NM for various end applications. This chapter provides a summary of numerous 2D-NM in terms of synthesis, characteristics (mechanical, electrical, and thermal properties), and their applications, primarily in supercapacitors (SCs) and solar cells. So far, several synthetic processes have been developed for creating 2D-NM using CVD and exfoliation techniques.

These processes enable easy control over the shape, size, crystal structure, and stability of the 2D-NM. The 2D-NM exhibit exceptional performance in various areas, particularly in SCs, solar cells, and electronic devices. By manipulating the properties of 2D-NM, we can effectively enhance its performance in various applications. The researcher's focus lies in improving the mechanical, electrical, and thermal properties. The data suggests that modifying the lattice structure and introducing metal/non-metal doping can enhance the structural stability, leading to improvements in thermal, mechanical, and electrical characteristics. Through these modifications, we can easily tailor the performance of 2D-NM for specific applications. The aforementioned discussion indicates that the unique properties of 2D-NM and its hybrid materials hold great potential for next-generation high-performing devices.

KEYWORDS

- **doping**
- **electrical stability**
- **electrochemical deposition**
- **graphene**
- **hybrid materials**
- **mechanical stability**
- **solar cells**
- **supercapacitors**
- **two-dimensional (2D)**

REFERENCES

1. Tiwari, J. N., Tiwari, R. N., & Kim, K. S., (2012). Zero-dimensional, one-dimensional, two-dimensional, and three-dimensional nanostructured materials for advanced electrochemical energy devices. *Progress in Materials Science, 57*(4), 724–803. https://doi.org/https://doi.org/10.1016/j.pmatsci.2011.08.003.
2. Rizvi, M., Gerengi, H., & Gupta, P., (2022). *Functionalization of Nanomaterials: Synthesis and Characterization, Functionalized Nanomaterials for Corrosion Mitigation: Synthesis, Characterization, and Applications* (pp. 1 26). American Chemical Society. https://doi.org/doi:10.1021/bk-2022-1418.ch001. 10.1021/bk-2022-1418.ch001.

3. Ashfaq, M., Talreja, N., Chauhan, D., Rodríguez, C. A., Mera, A. C., & Ramalinga, V. M., (2022). Synthesis of reduced graphene oxide incorporated bimetallic (Cu/Bi) nanorods based photocatalyst materials for the degradation of gallic acid and bacteria. *Journal of Industrial and Engineering Chemistry, 110*, 447–455. https://doi.org/https://doi.org/10.1016/j.jiec.2022.03.023.
4. Sun, Z., Talreja, N., Tao, H., Texter, J., Muhler, M., Strunk, J., & Chen, J., (2018). Catalysis of carbon dioxide photoreduction on nanosheets: Fundamentals and challenges. *Angewandte Chemie International Edition, 57*(26), 7610–7627. https://doi.org/https://doi.org/10.1002/anie.201710509.
5. Moraes, S. S., Tavallaie, R., Sandiford, L., Tilley, R. D., & Gooding, J. J., (2016). Gold coated magnetic nanoparticles: From preparation to surface modification for analytical and biomedical applications. *Chemical Communications, 52*(48), 7528–7540. https://doi.org/10.1039/C6CC03225G.
6. Gawande, M. B., Goswami, A., Felpin, F. X., Asefa, T., Huang, X., Silva, R., Zou, X., et al., (2016). Cu and Cu-based nanoparticles: Synthesis and applications in catalysis. *Chemical Reviews, 116*(6), 3722–3811. https://doi.org/10.1021/acs.chemrev.5b00482.
7. Jeevanandam, J., & Barhoum, A., (2018). *Review on Nanoparticles and Nanostructured Materials: History, Sources, Toxicity and Regulations,9*, 1050–1074. https://doi.org/10.3762/bjnano.9.98.
8. Sasidharan, V., Damiri, F., Talreja, N., Chauhan, D., Mangalaraja, R. V., Berrada, M., & Ashfaq, M., (2022). Carbon-based nanomaterials: An efficient tool for improving the nutritional quality of crops. In: Aftab, T., & Hakeem, K. R., (eds.), *Metabolic Engineering in Plants* (pp. 375–389). Springer Nature Singapore, Singapore. https://doi.org/10.1007/978-981-16-7262-0_17.
9. Tao, H., Gao, Y., Talreja, N., Guo, F., Texter, J., Yan, C., & Sun, Z., (2017). Two-dimensional nanosheets for electrocatalysis in energy generation and conversion. *Journal of Materials Chemistry A, 5*(16), 7257–7284. https://doi.org/10.1039/C7TA00075H.
10. Tiwari, S. K., Sahoo, S., Wang, N., & Huczko, A., (2020). Graphene research and their outputs: Status and prospect. *Journal of Science: Advanced Materials and Devices,5*(1), 10–29. https://doi.org/https://doi.org/10.1016/j.jsamd.2020.01.006.
11. Smith, A. T., LaChance, A. M., Zeng, S., Liu, B., & Sun, L., (2019). Synthesis, properties, and applications of graphene oxide/reduced graphene oxide and their nanocomposites. *NanoMaterials Science, 1*(1), 31–47. https://doi.org/https://doi.org/10.1016/j.nanoms.2019.02.004.
12. Mir, S. H., Yadav, V. K., & Singh, J. K., (2020). Recent advances in the carrier mobility of two-dimensional materials: A theoretical perspective. *ACS Omega, 5*(24), 14203–14211. https://doi.org/10.1021/acsomega.0c01676.
13. Peng, Z., Chen, X., Fan, Y., Srolovitz, D. J., & Lei, D., (2020). Strain engineering of 2D semiconductors and graphene: From strain fields to band-structure tuning and photonic applications. *Light: Science & Applications,9*(1), 190. https://doi.org/10.1038/s41377-020-00421-5.
14. Ashfaq, M., Talreja, N., Chauhan, D., & Viswanathan, M. R., (2022). Synthesis of Cu-doped 2D-WS_2 nanosheet-based nano-antibiotic materials for inhibiting *E. coli* and *S. aureus* bacterial strains. *New Journal of Chemistry, 46*(12), 5581–5587. https://doi.org/10.1039/D1NJ05931A.
15. Sultana, A., Omar, R. A., Talreja, N., Chauhan, D., Mangalaraja, R. V., & Ashfaq, M., (2022). Chapter 28–copper-based metal-organic framework for environmental

applications. In: Abd-Elsalam, E. A., (ed.), *Copper Nanostructures: Next-Generation of Agrochemicals for Sustainable Agroecosystems* (pp. 701–717). Elsevier. https://doi.org/https://doi.org/10.1016/B978-0-12-823833-2.00007-6.

16. Omar, R. A., Talreja, N., Chauhan, D., Mangalaraja, R. V., & Ashfaq, M., (2022). Chapter 14 – nano metal-carbon–based materials: Emerging platform for the growth and protection of crops. In: Balestra, G. M., & Fortunati, E., (eds.), *Nanotechnology-Based Sustainable Alternatives for the Management of Plant Diseases* (pp. 341–354). Elsevier. https://doi.org/https://doi.org/10.1016/B978-0-12-823394-8.00013-5.
17. Sasidharan, V., Sachan, D., Chauhan, D., Talreja, N., & Ashfaq, M., (2021). Three-dimensional (3D) polymer—Metal–carbon framework for efficient removal of chemical and biological contaminants. *Scientific Reports, 11*(1), 7708. https://doi.org/10.1038/s41598-021-86661-w.
18. Ahn, S., Kim, G., Nayak, P. K., Yoon, S. I., Lim, H., Shin, H. J., & Shin, H. S., (2016). Prevention of transition metal dichalcogenide photo degradation by encapsulation with h-BN layers. *ACS Nano, 10*(9), 8973–8979. https://doi.org/10.1021/acsnano.6b05042.
19. Suzuki, Y., & Watanabe, K., (2020). Excitons in two-dimensional atomic layer materials from time-dependent density functional theory: Mono-layer and bi-layer hexagonal boron nitride and transition-metal dichalcogenides. *Physical Chemistry Chemical Physics, 22*(5), 2908–2916. https://doi.org/10.1039/C9CP06034K.
20. Ashfaq, M., Wongpakham, T., Talreja, N., Chauhan, D., Tharasanit, T., & Srituravanich, W., (2022). Synthesis of polymeric composite grafted with mineral particles/graphene oxide-based biomaterial: A promising robust hemostatic bandage. *Materials Today Communications, 33*, 104786. https://doi.org/https://doi.org/10.1016/j.mtcomm.2022.104786.
21. Talreja, N., Ashfaq, M., Chauhan, D., & Mangalaraja, R. V., (2023). Cu-MXene: A potential biocide for the next-generation biomedical application. *Materials Chemistry and Physics, 294*, 127029. https://doi.org/https://doi.org/10.1016/j.matchemphys.2022.127029.
22. Bernardi, M., Palummo, M., & Grossman, J. C., (2013). Extraordinary sunlight absorption and one nanometer thick photovoltaic's using two-dimensional monolayer materials. *Nano Letters, 13*(8), 3664–3670. https://doi.org/10.1021/nl401544y.
23. Samy, O., Zeng, S., Birowosuto, M. D., & El Moutaouakil, E., (2021). A review on MoS_2 properties, synthesis, sensing applications and challenges. *Crystals, 11*(4). https://doi.org/10.3390/cryst11040355.
24. Singhal, J., Verma, S., & Kumar, S., (2022). The physio-chemical properties and applications of 2D nanomaterials in agricultural and environmental sustainability. *Science of The Total Environment, 837*, 155669. https://doi.org/https://doi.org/10.1016/j.scitotenv.2022.155669.
25. Deng, S., Sumant, A. V., & Berry, V., (2018). Strain engineering in two-dimensional nanomaterials beyond graphene. *Nano Today, 22*, 14–35. https://doi.org/https://doi.org/10.1016/j.nantod.2018.07.001.
26. Derakhshi, M., & Daemi, S., (2022). *Two-Dimensional Nanomaterials Beyond Graphene for Biomedical Applications, 13*(1). https://doi.org/10.3390/jfb13010027.
27. Malik, P., Gupta, R., Malik, V., & Ameta, R. K., (2021). Emerging nanomaterials for improved biosensing. *Measurement: Sensors, 16*, 100050. https://doi.org/https://doi.org/10.1016/j.measen.2021.100050.

28. Bolotsky, A., Butler, D., Dong, C., Gerace, K., Glavin, N. R., Muratore, C., Robinson, J. A., & Ebrahimi, A., (2019). Two-dimensional materials in biosensing and healthcare: From *in vitro* diagnostics to optogenetics and beyond. *ACS Nano, 13*(9), 9781–9810. https://doi.org/10.1021/acsnano.9b03632.
29. Li, X., Zhu, J., & Wei, B., (2016). Hybrid nanostructures of metal/two-dimensional nanomaterials for plasmon-enhanced applications. *Chemical Society Reviews, 45*(11), 3145–3187. https://doi.org/10.1039/C6CS00195E.
30. Yang, F., Song, P., Ruan, M., & Xu, W., (2019). Recent progress in two-dimensional nanomaterials: Synthesis, engineering, and applications. *FlatChem, 18*, 100133. https://doi.org/https://doi.org/10.1016/j.flatc.2019.100133.
31. Zhang, H., Fan, T., Chen, W., Li, Y., & Wang, B., (2020). Recent advances of two-dimensional materials in smart drug delivery nano-systems. *Bioactive Materials, 5*(4), 1071–1086. https://doi.org/10.1016/j.bioactmat.2020.06.012.
32. Huang, H., Jiang, R., Feng, Y., Ouyang, H., Zhou, N., Zhang, X., & Wei, Y., (2020). Recent development and prospects of surface modification and biomedical applications of MXenes. *Nanoscale, 12*(3), 1325–1338. https://doi.org/10.1039/c9nr07616f.
33. Jaffar, S. S., Saallah, S., Misson, M., Siddiquee, S., Roslan, J., Saalah, S., & Lenggoro, W., (2022). Recent development and environmental applications of nanocellulose-based membranes. *Membranes, 12*(3). https://doi.org/10.3390/membranes12030287.
34. Zhang, J., Chen, H., Zhao, M., Liu, G., & Wu, J., (2020). 2D nanomaterials for tissue engineering application. *Nano Research, 13*(8), 2019–2034. https://doi.org/10.1007/s12274-020-2835-4.
35. Hu, T., Mei, X., Wang, Y., Weng, X., Liang, R., & Wei, M., (2019). Two-dimensional nanomaterials: Fascinating materials in biomedical field. *Science Bulletin, 64*(22), 1707–1727. https://doi.org/https://doi.org/10.1016/j.scib.2019.09.021.
36. Li, L., Zhou, M., Jin, L., Liu, L., Mo, Y., Li, X., Mo, Z., et al., (2019). Research progress of the liquid-phase exfoliation and stable dispersion mechanism and method of graphene. *Frontiers in Materials, 6*. https://doi.org/10.3389/fmats.2019.00325.
37. Ashfaq, M., Talreja, N., Chauhan, D., Afreen, S., Sultana, A., & Srituravanich, W., (2022). Two-dimensional (2D) hybrid nanomaterials for diagnosis and treatment of cancer. *Journal of Drug Delivery Science and Technology, 70,* 103268. https://doi.org/https://doi.org/10.1016/j.jddst.2022.103268.
38. Georgakilas, V., Otyepka, M., Bourlinos, A. B., Chandra, V., Kim, N., Kemp, K. C., Hobza, P., et al., (2012). Functionalization of graphene: Covalent and non-covalent approaches, derivatives, and applications. *Chemical Reviews, 112*(11), 6156–6214. https://doi.org/10.1021/cr3000412.
39. Gómez-Santos, G., (2009). Thermal Van Der Waals interaction between graphene layers. *Physical Review B, 80*(24), 245424. https://doi.org/10.1103/PhysRevB.80.245424.
40. Guan, Z., Wang, C., Li, W., Luo, S., Yao, Y., Yu, S., Sun, R., & Wong, C. P., (2018). A facile and clean process for exfoliating MoS_2 nanosheets assisted by a surface-active agent in aqueous solution. *Nanotechnology, 29*(42), 425702. https://doi.org/10.1088/1361-6528/aad676.
41. Tan, X., Kang, W., Liu, J., & Zhang, C., (2019). Synergistic exfoliation of MoS_2 by ultrasound sonication in a supercritical fluid based complex solvent. *Nanoscale Research Letters, 14*(1), 317. https://doi.org/10.1186/s11671-019-3126-4.
42. Sahoo, D., Kumar, B., Sinha, J., Ghosh, S., Roy, S. S., & Kaviraj, B., (2020). Cost effective liquid phase exfoliation of MoS_2 nanosheets and photocatalytic activity for

wastewater treatment enforced by visible light. *Scientific Reports, 10*(1), 10759. https://doi.org/10.1038/s41598-020-67683-2.

43. Sankararamakrishnan, N., Chauhan, D., & Dwivedi, J., (2016). Synthesis of functionalized carbon nanotubes by floating catalytic chemical vapor deposition method and their sorption behavior toward arsenic. *Chemical Engineering Journal, 284*, 599–608. https://doi.org/https://doi.org/10.1016/j.cej.2015.08.145.
44. Ashfaq, M., Verma, N., & Khan, S., (2018). Novel polymeric composite grafted with metal nanoparticle-dispersed CNFs as a chemiresistive non-destructive fruit sensor material. *Materials Chemistry and Physics, 217*, 216–227. https://doi.org/https://doi.org/10.1016/j.matchemphys.2018.06.056.
45. Alam, A., Saeed, G., & Lim, S., (2020). One-step synthesis of 2D–2D $Co(OH)_2$–$MoSe_2$ hybrid nanosheets as an efficient electrode material for high-performance asymmetric supercapacitor. *Journal of Electro analytical Chemistry, 879*, 114775. https://doi.org/https://doi.org/10.1016/j.jelechem.2020.114775.
46. Si, A., Kyzas, G. Z., Pal, K., & De Souza, Jr. F.G., (2021). Graphene functionalized hybrid nanomaterials for industrial-scale applications: A systematic review. *Journal of Molecular Structure, 1239*, 130518. https://doi.org/https://doi.org/10.1016/j.molstruc.2021.130518.
47. Han, X., Huang, J., Lin, H., Wang, Z., Li, P., & Chen, Y., (2018). *2D Ultrathin MXene-Based Drug-Delivery Nanoplatform for Synergistic Photothermal Ablation and Chemotherapy of Cancer, 7*(9), e1701394. https://doi.org/10.1002/adhm.201701394.
48. Judek, J., Gertych, A. P., Świniarski, M., Łapińska, A., Dużyńska, A., & Zdrojek, M., (2015). High accuracy determination of the thermal properties of supported 2D materials. *Scientific Reports, 5*(1), 12422. https://doi.org/10.1038/srep12422.
49. Feng, X., Kulish, V. V., Wu, P., Liu, X., & Ang, K. W., (2016). Anomalously enhanced thermal stability of phosphorene via metal adatom doping: An experimental and first-principles study. *Nano Research, 9*(9), 2687–2695. https://doi.org/10.1007/s12274-016-1156-0.
50. Tang, W., Tang, S., Zhang, C., Ma, Q., Xiang, Q., Yang, Y. W., & Luo, J., (2018). Simultaneously enhancing the thermal stability, mechanical modulus, and electrochemical performance of solid polymer electrolytes by incorporating 2D sheets. *Advanced Energy Materials, 8*(24), 1800866. https://doi.org/https://doi.org/10.1002/aenm.201800866.
51. Krishnamoorthy, A., Baradwaj, N., Nakano, A., Kalia, R. K., & Vashishta, P., (2021). Lattice thermal transport in two-dimensional alloys and fractal heterostructures, *Scientific Reports, 11*(1), 1656. https://doi.org/10.1038/s41598-021-81055-4.
52. Tong, Z., Pecchia, A., Yam, C., Bao, H., Dumitrică, T., & Frauenheim, T., (2022). Significant increase of electron thermal conductivity in Dirac semimetal Beryllonitrene by doping beyond van hove singularity. *Advanced Functional Materials, 32*(17), 2111556. https://doi.org/https://doi.org/10.1002/adfm.202111556.
53. Moore, A. L., & Shi, L., (2014). Emerging challenges and materials for thermal management of electronics. *Materials Today, 17*(4), 163–174. https://doi.org/https://doi.org/10.1016/j.mattod.2014.04.003.
54. Ali, H. M., Arshad, A., Jabbal, M., & Verdin, P. G., (2018). Thermal management of electronics devices with PCMs filled pin-fin heat sinks: A comparison. *International Journal of Heat and Mass Transfer, 117*, 1199–1204. https://doi.org/https://doi.org/10.1016/j.ijheatmasstransfer.2017.10.065.
55. Zhang, Y. F., Han, D., Zhao, Y. H., & Bai, S. L., (2016). High-performance thermal interface materials consisting of vertically aligned graphene film and polymer. *Carbon, 109*, 552–557. https://doi.org/https://doi.org/10.1016/j.carbon.2016.08.051.

56. Khan, K., Tareen, A. K., Aslam, M., Mahmood, A., khan, Q., Zhang, Y., Ouyang, Z., et al., (2020). Going green with batteries and supercapacitor: Two dimensional materials and their nanocomposites-based energy storage applications. *Progress in Solid State Chemistry, 58*, 100254. https://doi.org/https://doi.org/10.1016/j.progsolidstchem.2019.100254.
57. Sreeprasad, T. S., Nguyen, P., Kim, N., & Berry, V., (2013). Controlled, defect-guided, metal-nanoparticle incorporation onto MoS_2 via chemical and microwave routes: Electrical, thermal, and structural properties. *Nano Letters, 13*(9), 4434–4441. https://doi.org/10.1021/nl402278y.
58. Yang, J., Luo, X., Zhang, S., & Chen, L., (2016). Investigation of magnetic and electronic properties of transition metal doped Sc_2CT_2(T = O, OH or F) using a first principles study. *Physical Chemistry Chemical Physics, 18*(18), 12914–12919. https://doi.org/10.1039/C6CP00138F.
59. Xiong, K., Wang, P., Yang, G., Liu, Z., Zhang, H., Jin, S., & Xu, X., (2017). Functional group effects on the photo electronic properties of MXene (Sc_2CT_2, T = O, F, OH) and their possible photocatalytic activities. *Scientific Reports*, 15095.
60. Vargas-Bernal, R., (2019). Electrical properties of two-dimensional materials used in gas sensors. *Sensors, 19*(6). https://doi.org/10.3390/s19061295.
61. Postorino, S., Grassano, D., D'Alessandro, M., Pianetti, A., Pulci, O., & Palummo, M., (2020). Strain-induced effects on the electronic properties of 2D materials. *Nanomaterials and Nanotechnology, 10*, 1847980420902569. https://doi.org/10.1177/ 1847980420902569.
62. Dong, Z., Cui, H., Zhang, H., Wang, F., Zhan, X., Mayer, F., Nestler, B., et al., (2021). 3D printing of inherently nanoporous polymers via polymerization-induced phase separation. *Nature Communications, 12*(1), 247. https://doi.org/10.1038/s41467-020-20498-1.
63. Castro-Gutiérrez, J., Celzard, A., & Fierro, V., (2020). Energy storage in supercapacitors: Focus on tannin-derived carbon electrodes. *Frontiers in Materials, 7*.
64. Pomerantseva, E., & Gogotsi, Y., (2017). Two-dimensional heterostructures for energy storage. *Nature Energy, 2*(7), 17089. https://doi.org/10.1038/nenergy.2017.89.
65. Krishnamoorthy, D., & Prakasam, A., (2021). Graphene hybridized with tungsten disulfide (WS_2) based heterojunctions photo anode materials for high performance dye sensitized solar cell device (DSSCs) applications. *Journal of Cluster Science, 32*(3), 621–630. https://doi.org/10.1007/s10876-020-01828-1.
66. Zhang, X., Xu, Y., Zhang, J., Dong, S., Shen, L., Gupta, A., & Bao, N., (2018). Synthesis of wurtzite Cu_2ZnSnS_4 nanosheets with exposed high-energy (002) facets for fabrication of efficient pt-free solar cell counter electrodes. *Scientific Reports, 8*(1), 248. https://doi.org/10.1038/s41598-017-18631-0.
67. Fang, X., Song, T., Liu, R., & Sun, B., (2014). Two-dimensional CoS nanosheets used for high-performance organic–inorganic hybrid solar cells. *The Journal of Physical Chemistry C, 118*(35), 20238–20245. https://doi.org/10.1021/jp506345a.
68. Mahesh, A., (2011). Photovoltaic performance of ZnO nanosheets solar cell sensitized with beta-substituted porphyrins. *Journal of Nanomaterials, 2011*, 301873. https://doi.org/10.1155/2011/301873.
69. Jiang, T., Yin, N., Bai, Z., Dai, P., Yu, X., Wu, M., & Li, G., (2018). Wet chemical synthesis of S doped Co_3O_4 nanosheets/reduced graphene oxide and their application in dye sensitized solar cells. *Applied Surface Science, 450*, 219–227. https://doi.org/https://doi.org/10.1016/j.apsusc.2018.04.148.
70. Huang, S., He, Q., Chen, W., Qiao, Q., Zai, J., & Qian, X., (2015). Ultrathin $FeSe_2$ nanosheets: Controlled synthesis and application as a heterogeneous catalyst in

dye-sensitized solar cells. *Chemistry – A European Journal, 21*(10), 4085–4091. https://doi.org/https://doi.org/10.1002/chem.201406124.

71. Sheng, L., Liao, T., Kou, L., & Sun, Z., (2017). Single-crystalline ultrathin 2D TiO_2 nanosheets: A bridge towards superior photovoltaic devices. *Materials Today Energy, 3*, 32–39. https://doi.org/https://doi.org/10.1016/j.mtener.2016.12.004.
72. Dharamalingam, K., Arjun, K. B., Ramalingam, G., Sasi, F. S., Raju, K., Senthil, K. P., Govindaraju, S., & Thangavel, E., (2022). The role of sodium dodecyl sulfate mediated hydrothermal synthesis of MoS_2 nanosheets for photocatalytic dye degradation and dye-sensitized solar cell application. *Chemosphere, 294*, 133725. https://doi.org/https://doi.org/10.1016/j.chemosphere.2022.133725.
73. Jiang, T., Yang, S., Dai, P., Yu, X., Bai, Z., Wu, M., Li, G., & Tu, C., (2018). Economic synthesis of Co_3S_4 ultrathin nanosheet/reduced graphene oxide composites and their application as an efficient counter electrode for dye-sensitized solar cells. *Electrochimica Acta, 261*, 143–150. https://doi.org/https://doi.org/10.1016/j.electacta.2017.12.121.
74. Afzali, M., Mostafavi, A., & Shamspur, T., (2020). Improved perovskite solar cell with 2H–TaS_2 nanosheets as an electron transport layer using microwave irradiation. *Journal of Alloys and Compounds, 817*, 152742. https://doi.org/https://doi.org/10.1016/j.jallcom.2019.152742.
75. Huang, K. J., Wang, L., Liu, Y. J., Wang, H. B., Liu, Y. M., & Wang, L. L., (2013). Synthesis of polyaniline/2-dimensional graphene analog MoS_2 composites for high-performance supercapacitor. *Electrochimica Acta, 109*, 587–594. https://doi.org/https://doi.org/10.1016/j.electacta.2013.07.168.
76. Quan, H., Cheng, B., Xiao, Y., & Lei, S., (2016). One-pot synthesis of α-Fe_2O_3 nanoplates-reduced graphene oxide composites for supercapacitor application. *Chemical Engineering Journal, 286,* 165–173. https://doi.org/https://doi.org/10.1016/j.cej.2015.10.068.
77. Ansari, S. A., Fouad, H., Ansari, S. G., Sk, M. P., & Cho, M. H., (2017). Mechanically exfoliated MoS_2 sheet coupled with conductive polyaniline as a superior supercapacitor electrode material. *Journal of Colloid and Interface Science, 504*, 276–282. https://doi.org/https://doi.org/10.1016/j.jcis.2017.05.064.
78. Navarro-Suárez, A. M., Maleski, K., Makaryan, T., Yan, J., Anasori, B., & Gogotsi, Y., (2018). 2D titanium carbide/reduced graphene oxide heterostructures for supercapacitor applications. *Batteries & Supercaps, 1*(1), 33–38. https://doi.org/https://doi.org/10.1002/batt.201800014.
79. Rajendran, K., Nagamony, P., & Annal, T. H., (2022). Template-free synthesis of 2D-pristine VS_4 nanosheets for supercapacitor applications. *Physica Status Solidi (a)* n/a(n/a), 2200203. https://doi.org/https://doi.org/10.1002/pssa.202200203.
80. Chang, A., Zhang, C., Yu, Y., Yu, Y., & Zhang, B., (2018). Plasma-assisted synthesis of $NiSe_2$ ultrathin porous nanosheets with selenium vacancies for supercapacitor. *ACS Applied Materials & Interfaces, 10*(49), 41861–41865. https://doi.org/10.1021/acsami.8b16072.
81. Yuan, W., Cheng, L., Wu, H., Zhang, Y., Lv, S., & Guo, X., (2018). One-step synthesis of 2D-layered carbon wrapped transition metal nitrides from transition metal carbides (MXenes) for supercapacitors with ultrahigh cycling stability. *Chemical Communications, 54*(22), 2755–2758. https://doi.org/10.1039/C7CC09017J.
82. Geng, Q., Huang, G., Liu, Y., Li, Y., Liu, L., Yang, X., Wang, Q., & Zhang, C., (2019). Facile synthesis of B/N co-doped 2D porous carbon nanosheets derived from ammonium humate for supercapacitor electrodes. *Electrochimica Acta, 298*, 1–13. https://doi.org/https://doi.org/10.1016/j.electacta.2018.12.038.

83. Zhang, Z., Huang, X., Lu, W., Qiu, X., Ma, T., & Xia, W., (2020). Synthesis of 2D layered Nb_2SnC at low sintering temperature and its application for high-performance supercapacitors. *Journal of Solid-State Chemistry, 288,* 121425. https://doi.org/https://doi.org/10.1016/j.jssc.2020.121425.
84. Naz, R., Imtiaz, M., Liu, Q., Yao, L., Abbas, W., Li, T., Zada, I., et al., (2019). Highly defective 1T-MoS_2 nanosheets on 3D reduced graphene oxide networks for supercapacitors. *Carbon, 152*, 697–703. https://doi.org/https://doi.org/10.1016/j.carbon.2019.06.009.
85. Venkateshalu, S., Cherusseri, J., Karnan, M., Kumar, K. S., Kollu, P., Sathish, M., Thomas, J., et al., (2020). New method for the synthesis of 2D vanadium nitride (MXene) and its application as a supercapacitor electrode. *ACS Omega, 5*(29), 17983–17992. https://doi.org/10.1021/acsomega.0c01215.

CHAPTER 3

Bismuth Ferrite/Barium Titanate ($BiFeO_3$/$BaTiO_3$) Thick Films for High Energy Applications: A Review

DHANYA RAJ, T. V. VENKIDESH, and M. V. MANJUSHA

Department of Physics, The Cochin College, Kochi, Ernakulam, Kerala, India

ABSTRACT

Previous studies have primarily focused on lead-based fuels. Research over the past decades has shown that lead-containing ferroelectric components possess high energy storage densities and high energy efficiencies. However, these components consist of solids with complex compositions, making them very toxic and harmful to the environment. Subsequent research has been mainly focused on finding alternative materials that are easy to manufacture, environmentally friendly, and cost-effective. A new material is needed to overcome these limitations. Researchers have investigated single-component lead-free perovskites for their application in thick film dielectric capacitors, highlighting the importance of nanotechnology. Among these, Barium Titanate ($BaTiO_3$) and bismuth ferrite ($BiFeO_3$) are the most well-known thick films, both exhibiting excellent dielectric properties. An ideal material for highly efficient energy storage devices would have a high dielectric breakdown electric field, excellent impact resistance, low permanent polarization, and high saturation polarization. $BiFeO_3$ exhibits a largely saturated polarization, allowing it to be tailored in terms of electrical properties via strain or chemical doping. Barium titanate has excellent dielectric properties. Its relative dielectric constant is very large, and the value of the breakdown field is large, hence it is extensively used in multi-layer ceramic capacitors (MLCC). The development of lead-free

Sustainability in Energy and Environment: Engineered Materials and Smart Computational Techniques.
Ann Rose Abraham, Heru Susanto, A. K. Haghi, & Kian Harırı Aslı (Eds.)

dielectric capacitors with high energy storage density and efficiency can be highly promoted using $BiFeO_3$/$BaTiO_3$ bi-layer thick films, and this review discusses the progress in this field. Bi-layer thick films are deposited onto Pt/Ti/SiO_2/(100) Si substrate using $LaNiO_3$ buffer. As a result of this thick film's low remnant polarization and largely saturated polarization, it exhibits a slim polarization electric field hysteresis loop that enhances its energy storage capability. Many studies on bi-layer films have shown that they have excellent energy storage properties. In this researched field, many papers have come out on how to effectively use $BiFeO_3$-Barium Titanate bi-layer thick films, this chapter aims to summarize the reading of these papers.

3.1 INTRODUCTION

Energy is a very important and necessary factor in our daily life. In recent decades, the world's population has grown, and so have people's needs. Population growth, massive climate change, depletion of fossil fuels, and pollution of natural resources require the development of environmentally friendly and clean renewable energy sources. Reducing CO_2 emissions and fossil fuel use requires more reliance on renewable and sustainable energy sources such as solar, wind, tidal, and hydropower, which are only intermittently available [1]. Energy storage and energy conversion technology are the biggest challenges of this century. They also open up promising possibilities for the construction and development of energy storage systems. Electrochemical energy storage (EES) plays an important role in the available energy storage technologies. Electrochemical capacitors and rechargeable batteries are the most promising energy storage devices. They have very high energy production and power density, and excellent cycle stability [2].

EES technology includes lead-acid batteries. Lead-acid battery technology has long been used for energy storage. They are very cheap to manufacture and easy to maintain, but they pollute the environment and have very low energy density and life expectancy [3]. The advent of lithium-ion batteries brought about a major shift in energy storage. Their main advantages are high energy density, self-discharge properties, low maintenance, and favorable load characteristics [4]. However, the production cost is very high, and ageing is also a drawback. Studies over the past decades have shown that lead-containing ferroelectric films have very high energy efficiency and energy storage density. However, increasing environmental concerns from their use have led to a search for storage devices that are suitable for manufacturing by reducing their environmental impact [5].

Recently, researchers worldwide have been working on developing energy storage devices that are environmentally friendly, easy to manufacture, and lead-free. This is where nanotechnology applications come into play [6]. Nanomaterials (NMs) offer improved ionic transport and electronic conductivity compared to existing batteries and supercapacitors (SCs) that use NMs [7]. Electroceramics, a class of materials with unique electrical, optical, and magnetic properties, find applications in communication, power conversion, energy storage, and various electronic applications [8]. Dielectrics, ferroelectrics, piezoelectrics, pyroelectrics, magnetic ceramics, and ceramic ionic conductors are all types of electroceramics. Dielectric capacitors possess a high electrostatic storage capacity and can release stored energy rapidly [9]. The most commonly used materials for dielectric applications are Barium Titanate (BT – $BaTiO_3$), Lead Zirconate Titanate (PZT – $Pb(Zr, Ti)O_3$), and bismuth ferrite (BF – $BiFeO_3$). Among these, BT is used for manufacturing multilayer ceramic capacitors, PZT is employed in pyro, piezo, and ferroelectric applications, and BF is utilized in ferroelectric relaxor applications [10].

$BiFeO_3$/Barium Titanate ceramic is an electroceramic developed to replace Lead Zirconate Titanate. Many publications have shown that the $BiFeO_3/BaTiO_3$ system is a highly reliable and efficient ceramic energy storage system. This review focuses on the study of energy storage properties applied to $BiFeO_3/BaTiO_3$ bi-layer thick films, their applications, and new developments. In this study, it was found that the $BiFeO_3/BaTiO_3$ bi-layer thick films have improved energy density and charge-discharge efficiency superior to single-layer thin films of this material. Therefore, the $BiFeO_3/BaTiO_3$ bi-layer thick film is a great innovation in the fabrication of lead-free energy storage devices.

3.2 HIGH ENERGY APPLICATIONS OF BFO/BTO

Nanoscience and nanotechnology are growing areas of science. The rapid development of nanoscience and nanotechnology allows us to expand into all industries. The possibilities it brings are not limited to any particular field [11]. Today, nanotechnology is successfully used in various sectors, and energy storage is a key area of application. Lele Peng et al. have proposed rational material structure design using nanotechnology to enhance the overall performance of electronic energy storage devices [12]. Nanomaterials with particle sizes in the hundreds of nanometer range offer excellent ion storage specific surface area (SSA), ion exchange, surface redox reactions, and shorter ion transport distances compared to bulk materials. These characteristics improve attributes such as capacity and rate capability. As a 2D nanomaterial, this material exhibits excellent charge storage capability due to its extremely high SSA and high aspect ratio [13].

Dielectrics are materials with very high electrical resistance. Their specific electrical resistance is close to 108 Ωm. The formation or reorientation of an electric field creates lattice polarization that can be used to store electrical energy [14]. When a dielectric is placed under an electric field, the electric field is developed by the polarization of the atoms and molecules in the material, which oppose the applied electric field. For these reasons, dielectric capacitors can deliver charge very quickly [15]. This distinguishes dielectric materials from conventional batteries. In conventional batteries, the release of charge is slow due to the use of chemical reactions. On the other hand, dielectrics have a longer life compared to conventional batteries, and the short life of conventional batteries is due to irreversible chemical reactions [16]. The energy density stored or released in a dielectric can be determined from its characteristic dielectric displacement versus the electric field curve (Figure 3.1) [17].

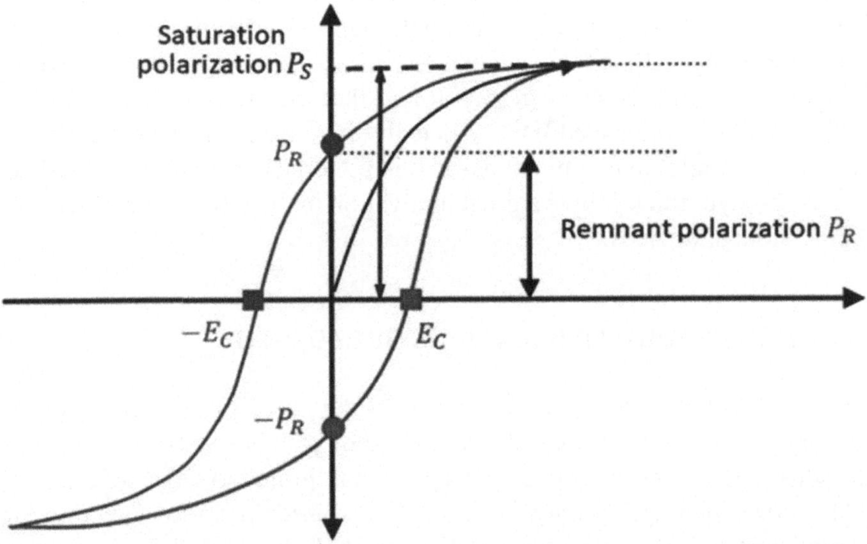

FIGURE 3.1 Schematic illustration of a ferroelectric hysteresis loop showing the total polarization in a ferroelectric material.

Source: Reprinted from Ref. [18].

The stored or charged energy density is W_c and is given by:

$$W_c = \int_0^{D_s} E\,dD \quad (1)$$

The released or discharged energy density will be W_d, and it is given by:

$$W_d = \int_{D_r}^{D_s} E\,dD \quad (2)$$

where; D_s is the saturated or maximum electrical displacement; and D_r is the residual electrical displacement.

This gives you energy efficiency:

$$\eta = \frac{W_d}{W_c} \quad (3)$$

For an ideal linear dielectric with relative permittivity ε_r:

$$D = \varepsilon_0 \varepsilon_r E \text{ and } D_r = 0 \quad (4)$$

Energy density is:

$$W_c = \frac{1}{2}\varepsilon_0 \varepsilon_r E_{max}^2 \quad (5)$$

$$W_d = W_c = \frac{1}{2}\varepsilon_0 \varepsilon_r E_{max}^2 \quad (6)$$

Efficiency $\eta = 1$

where; ε_0 is the vacuum permittivity; and E_{max} is the maximum applicable electric field, which increases with the ideal breakdown field E_b. For an ideal dielectric $E_{max} \approx E_b$. For nonlinear dielectrics, such as ferroelectrics, we usually do not have $D \propto E$ or $D_r = 0$, so $W_c > W_d$. Therefore, $\eta < 1$ [19]. However, Eqn. (5) can be used to estimate the energy density W_c of nonlinear dielectrics.

The polarization is given by:

$$P = \varepsilon_0 \varepsilon_r E \quad (7)$$

So, for dielectric ceramic materials to achieve good electrical density, the material must have a high ε_r, a low remanent polarization P_r, a high saturation polarization P_s and higher breakdown strength (BDS). A key determinant of the electrical density properties of a good electrical energy storage (EES) system is the material's BDS [20].

$BiFeO_3$ is one of the most widely studied multiferroic magnetoelectric materials, rich in dielectric and ferroelectric properties. Research into ceramic capacitors made of $BiFeO_3$ is increasing day by day [21]. Kotnala et al. proposed that it is mainly used for energy storage applications due to its high polarization and ferroelectric properties. Its Curie temperature is very high, around T_c = 1,043 K, and its Neel temperature is around T_n = 647 K. Due to the high Curie temperature, the ferroelectricity of this material is not lost even at high temperatures. Energy storage applications can be used for spintronics applications at room temperature. $BiFeO_3$ is different from other materials for several reasons. At room temperature, $BiFeO_3$ has a rhombohedral structure with point group R3C ($\alpha = \beta = \gamma = 59.4°$, a = b = c = 5). This material can achieve ferroelectricity due to the 6s lone pair of Bi and magnetic ordering due to the partially filled d orbitals of Fe (Figure 3.2) [22].

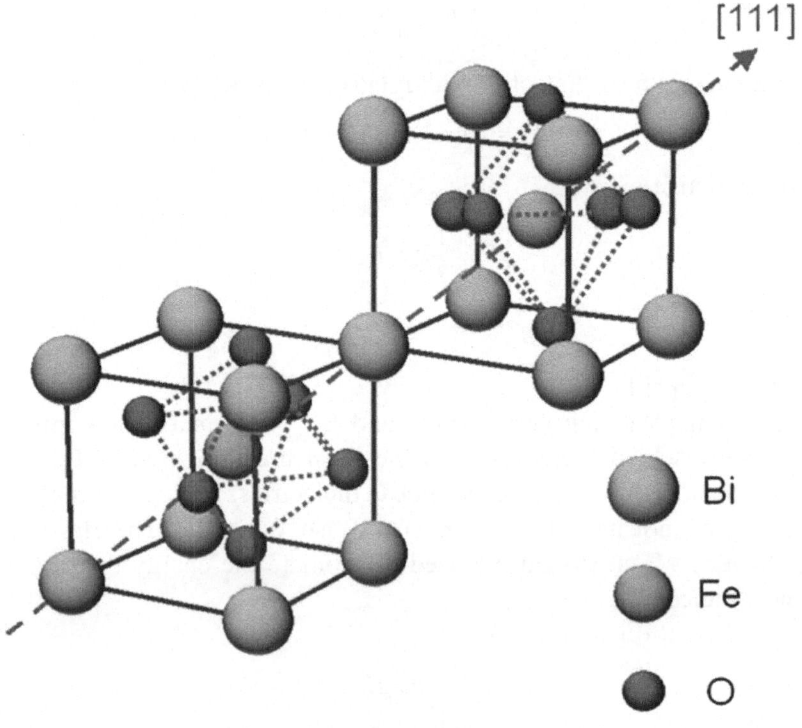

FIGURE 3.2 Schematic view of the R3C structure built up from two cubic perovskite $BiFeO_3$ unit cells.

Source: Reprinted with permission from Ref. [23].

Barium Titanate ($BaTiO_3$) is a material that is primarily used in the electronic device and ceramics industries. Its ferroelectric, thermoelectric, and piezoelectric properties make it a useful material in these areas [24]. Barium Titanate is a very strong dielectric material with a perovskite structure and is used in the electronics industry for the production of transducers, actuators, capacitors, and high-k dielectrics [25]. Due to its high dielectric constant, ceramic dielectric capacitors, capacitors, and ferroelectric memories can be made [26]. These uses make Barium Titanate an attractive material. Barium Titanate has a large band gap (~3.3 eV), which is larger than the energy corresponding to the wavelength of 405 nm [27]. If the optical band gap can be reduced, this material can be effectively used for ferroelectric photovoltaic effect (FEPV) applications in the visible region while maintaining ferroelectric properties [28]. The multilayer structure exhibits superior ferroelectric properties compared to the single Barium Titanate starting material [29].

The launch of the Bismuth Ferrite-Barium Titanate (BF-BT) thick film system is not limited to piezoelectric applications, but aims to provide a lead-free system that is equally applicable to all applications. It serves as an alternative to PZT [31]. Dawei Wang et al. suggested that $BaTiO_3$ and $BiFeO_3$ each have a T&R structure at room temperature, with the latter having a high critical temperature of about Tc = 825°C. In the BF-BT system, multiple ion species occupy both the A and B sites, in contrast to only the B sites in PZT (Figure 3.3) [10].

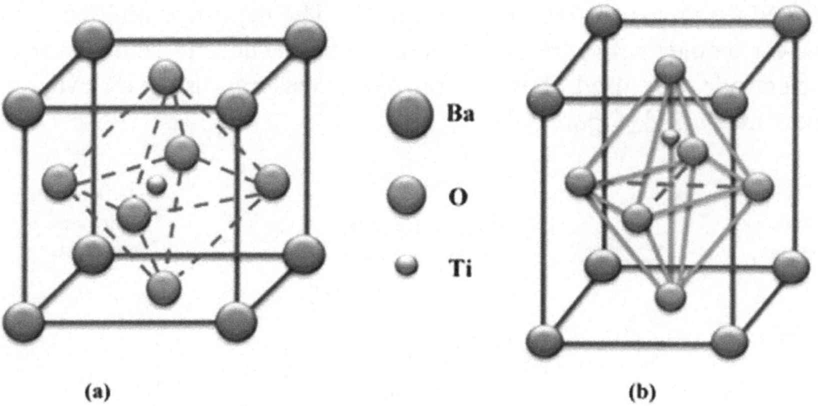

FIGURE 3.3 Schematic illustration of the perovskite structure of $BaTiO_3$. (a) Cubic lattice (above Curie temperature, >120°C); and (b) tetragonal lattice (above Curie temperature, <120°C).

Source: Reprinted with permission from Ref. [30]. Copyright © 2021, The Author(s), under exclusive licence to Springer Science Business Media, LLC, part of Springer Nature.

In the BF-BT system, Ba acts as a large blocking ion that facilitates the displacement of the Bi lone pair species. This system is ideal for high energy density storage, electrocaloric, and pyroelectric applications. Consequently, the BF-BT system shows great promise as an effective lead-free ceramic that can be utilized for centuries [32].

Studies have demonstrated that BF-BT films can be employed in the design of ferroelectric random-access memories (FeRAM) and microelectromechanical systems (MEMS). The temperature dependence of εr between the films and the bulk of the same composition has been thoroughly investigated [33]. Notably, the dielectric anomaly of the film is wider than that of the bulk due to the higher volume fraction of grain boundaries compared to the bulk and the difference in thermal expansion between the foil and the substrate [34]. Furthermore, altering the number of layers in the BF-BT film system (monolayer, bilayer, multilayer) significantly affects the dielectric and energy storage properties [35].

Liu Menglin et al. fabricated $BiFeO_3/BaTiO_3$ thick bilayer films with a thickness of about 1 μm. They fabricated this bilayer thick film using the magnetron sputtering method. The $BiFeO_3/BaTiO_3$ bilayer thick film was fabricated on a $Pt/Ti/SiO_2/(100)$ Si substrate buffered with a $LaNiO_3$ layer. The ceramic targets used in the deposition process were $BaTiO_3$, $Bi1.05FeO_3$, and $LaNiO_3$. The deposition was carried out in a multi-target magnetic sputtering system. The $LaTiO_3$, $BaTiO_3$, and $BiFeO_3$ layers were sequentially sputtered using a gas mixture of Ar and O_2. The deposited multilayer thick films were cooled with pure oxygen. For electrical characterization, a circular top electrode was used. Temperature was measured using an evaporated shadow mask [36] (Figure 3.4).

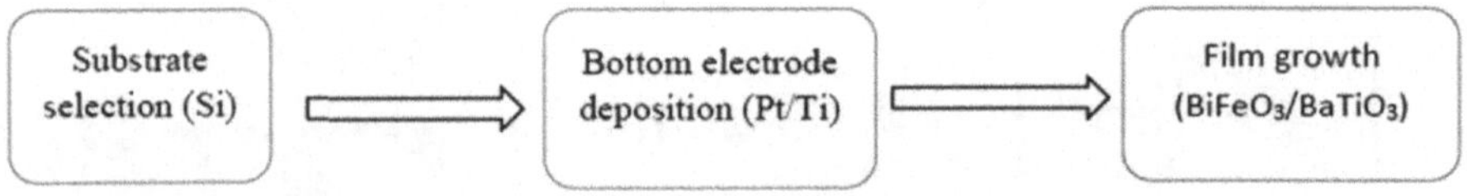

FIGURE 3.4 Schematics of experimental procedure.

X-ray diffraction (XRD) analysis was performed to determine the crystal orientation and phase structure characteristics of the $BiFeO_3/BaTiO_3$ double-layer thick film [37]. Field-emission scanning electron microscopy (SEM) and scanning probe microscopy were used to determine the cross-sectional morphology of the film. Scanning probe microscopy was used to examine the surface morphology, and its dielectric properties were determined by Menglin Liu et al. [36]. It was conducted using a high-precision digital bridge. The ferroelectric hysteresis loop (P-V) and leakage current (I-V) of the bilayer film at room temperature were determined using a Radiant

Precision Premium ferroelectric tester. XRD analysis was performed to study the microstructure and crystallographic orientation of the bi-layer thick films. The analysis showed that the film is dominated by the preferred (001) peak of the bulk perovskite structure and does not exhibit any crystals or secondary impurity phase. Menglin Liu et al. confirmed that the presence of the (100)-$LaNiO_3$ buffer layer facilitated the growth of the (001) oriented $BiFeO_3/BaTiO_3$ bilayer film. XRD investigations revealed that all the grains in the upper $BiFeO_3$ layer were densely packed with an average size of 200–300 nm.

Liu Menglin et al. were able to determine the energy storage density W_c and energy efficiency of Au/$BiFeO_3$/$BaTiO_3$/$LaNiO_3$/Pt/T capacitors from the hysteresis loop of the polarization electric field (P-E). They found a large maximum polarization value of about $P_m \sim 65$ μC/cm^2 and a small remanent polarization value of $P_r \sim 7.8$ μC/cm^2 for the bi-layer membrane, compared to a reduced field coercive field $E_c \sim 152$ kV/cm^2. From these characteristics, the film is known to have excellent energy storage properties [36].

Compared with pure $BaTiO_3$ deposited under the same conditions, we find that the energy storage density, W_c, of the bi-layer film reaches 71 J/cm^3. There is an improvement from 100% to 101%, but the energy efficiency remains at the same level (60% to 70%). This improvement in Wc is attributed to a significantly improved maximum polarization and a high dielectric strength comparable to that of pure $BaTiO_3$ films [39]. It was found that the "slimming" of the P-E loops in bilayer films was due to competition between space-charge effects and charge coupling between the layers. Therefore, the improved energy storage ability of $BiFeO_3/BaTiO_3$ bi-layer films may be attributed to the combination of the space-charge effect and charge-coupling effect between the layers, which is dominant at low electric fields [40]. The interfacial energy barrier between the $BiFeO_3$ and $BaTiO_3$ layers reduces the leakage current and exposes the films to large electric fields, thus making it possible to achieve high energy densities in the $BiFeO_3/BaTiO_3$ bi-layer films [41].

3.3 CONCLUSION

Many studies are currently being conducted on the need for $BiFeO_3/BaTiO_3$ bilayer lead-free thick films in energy storage applications. We wanted to review these studies. In comparison to the $BiFeO_3/BaTiO_3$ monolayer, we have found that the bilayer exhibits significantly better energy charge density and energy efficiency. This improvement is attributed to the combined effect of interlayer space charge and interlayer charge coupling. Additionally, these properties tend to increase with the number of layers. The dielectric constant Er is 33% higher than that of the pure $BaTiO_3$ film and 70% higher than that of the $BiFeO_3$ film. Furthermore, the leakage current is very low. As a result, $BiFeO_3/BaTiO_3$ bilayer films demonstrate excellent and efficient dielectric performance and energy storage properties. This makes them promising materials for lead-free thick film capacitor applications, as well as for electrocaloric and pyroelectric applications.

KEYWORDS

- **barium titanate**
- **bilayer**
- **bismuth ferrite**
- **capacitor**
- **energy storage**
- **ferroelectric tester**
- **high energy applications**
- **pyroelectric applications**
- **thick film**

REFERENCES

1. Wang, F., et al., (2021). Technologies and perspectives for achieving carbon neutrality. *The Innovation*, *2*(4), 100180. doi: 10.1016/j.xinn.2021.100180.
2. Kabeyi, M. J. B., & Olanrewaju, O. A., (2022). Sustainable energy transition for renewable and low carbon grid electricity generation and supply. *Front. Energy Res.*, *9*. [Online]. Available: https://www.frontiersin.org/articles/10.3389/fenrg.2021.743114 (accessed on 24 February 2024).
3. Challenges and progresses of energy storage technology and its application in power systems. *Journal of Modern Power Systems and Clean Energy*. [Online]. Available: https://link.springer.com/article/10.1007/s40565-016-0248-x (accessed on 24 February 2024).
4. Wu, F. B., Yang, B., & Ye, J. L., (2019). Chapter 2–technologies of energy storage systems. In: *Grid-scale Energy Storage Systems and Applications* (pp. 17–56). Academic Press. doi: 10.1016/B978-0-12-815292-8.00002-2.
5. Lead-carbon batteries toward future energy storage: From mechanism and materials to applications. *Electrochemical Energy Reviews*. [Online]. Available: https://link.springer.com/article/10.1007/s41918-022-00134-w (accessed on 24 February 2024).

6. *Overview of Current Development in Electrical Energy Storage Technologies and the Application Potential in Power System Operation*–ScienceDirect. [Online]. Available: https://www.sciencedirect.com/science/article/pii/S0306261914010290 (accessed on 24 February 2024).
7. *Nanomaterials for Energy Storage Applications*. [Online]. Available: https://www.researchgate.net/publication/352896211_Nanomaterials_for_Energy_Storage_Applications (accessed on 24 February 2024).
8. Mahmud, M. Z. A., (2023). A concise review of nanoparticles utilized energy storage and conservation. *J. Nanomater., 2023*, e5432099. doi: 10.1155/2023/5432099.
9. Cigolotti, V., Genovese, M., Piraino, F., & Fragiacomo, P., (2023). Applications–stationary | stationary energy storage system: Overview. In: *Reference Module in Chemistry, Molecular Sciences and Chemical Engineering* (p. B9780323960229000918). Elsevier. doi: 10.1016/B978-0-323-96022-9.00091-8.
10. Wang, D., et al., (2018). $BiFeO_3$ -$BaTiO_3$: A new generation of lead-free electroceramics. *J. Adv. Dielectr., 8*(6), 1830004. doi: 10.1142/S2010135X18300049.
11. Bayda, S., Adeel, M., Tuccinardi, T., Cordani, M., & Rizzolio, F., (2019). The history of nanoscience and nanotechnology: From chemical–physical applications to nanomedicine. *Molecules, 25*(1), 112. doi: 10.3390/molecules25010112.
12. Peng, L., Zhu, Y., & Yu, G., (2015). (Invited) hybrid two-dimensional nanosheet materials for energy storage devices. *ECS Meet. Abstr. MA, 2*(31), 1145. doi: 10.1149/MA2015-02/31/1145.
13. Kukkar, D., Vellingiri, K., Kumar, V., Deep, A., & Kim, K. H., (2018). A critical review on the metal sensing capabilities of optically active nanomaterials: Limiting factors, mechanism, and performance evaluation. *TrAC Trends Anal. Chem., 109*, 227–246. doi: 10.1016/j.trac.2018.09.009.
14. Veerapandiyan, V., Benes, F., Gindel, T., & Deluca, M., (2020). Strategies to improve the energy storage properties of perovskite lead-free relaxor ferroelectrics: A review. *Materials, 13*(24), 5742. doi: 10.3390/ma13245742.
15. Fletcher, N. H., Hilton, A. D., & Ricketts, B. W., (1996). Optimization of energy storage density in ceramic capacitors. *J. Phys. Appl. Phys., 29*(1), 253. doi: 10.1088/ 0022-3727/29/1/037.
16. Winter, M., & Brodd, R. J., (2004). What are batteries, fuel cells, and supercapacitors? *Chem. Rev., 104*(10), 4245–4270. doi: 10.1021/cr020730k.
17. Hao, X., Wang, Y., Yang, J., An, S., & Xu, J., (2012). High energy-storage performance in $Pb_{0.91}La_{0.09}(Ti_{0.65}Zr_{0.35})O_3$ relaxor ferroelectric thin films. *J. Appl. Phys., 112*(11), 114111. doi: 10.1063/1.4768461.
18. *Ferroelectrics & Ferroelectric Materials*. [Online]. Available: https://www.globalsino.com/EM/page3544.html (accessed on 24 February 2024).
19. Wang, Y., Zhou, X., Chen, Q., Chu, B., & Zhang, Q., (2010). Recent development of high energy density polymers for dielectric capacitors. *IEEE Trans. Dielectr. Electr. Insul., 17*(4), 1036–1042. doi: 10.1109/TDEI.2010.5539672.
20. *High-Temperature Dielectric Materials for Electrical Energy Storage* | Request PDF. [Online]. Available: https://www.researchgate.net/publication/323462799_High-Temperature_Dielectric_Materials_for_Electrical_Energy_Storage (accessed on 24 February 2024).
21. Han, S., et al., (2010). Synthesis and characterization of multiferroic $BiFeO_3$ powders fabricated by hydrothermal method. *Ceram. Int. – CERAM INT, 36*, 1365–1372. doi: 10.1016/j.ceramint.2010.01.020.

22. St. Peterburg State University, 7/9, Universitetskaya NAB., 199034 St. Petersburg, Russia et al., (2016). Magnetic and electrical characteristics of bismuth ferrite, depending on the impurities, method of preparation and size of the nanoparticles. *J. Nano- Electron. Phys.*, *8*(3), 03008-1-03008–4. doi: 10.21272/jnep.8(3).03008.
23. Shima, H., Naganuma, H., & Okamur, S., (2013). Optical properties of multiferroic $BiFeO_3$ films. In: Mastai, Y., (ed.), *Materials Science – Advanced Topics*. In-tech. doi: 10.5772/54908.
24. Kholodkova, A. A., et al., (2018). Properties of barium titanate ceramics based on powder synthesized in supercritical water. *Ceram. Int.*, *44*(11), 13129–13138. doi: 10.1016/j.ceramint.2018.04.135.
25. Raj, D., & Manjusha, M. V., (2023). Synthesis and optical characterization of Fe doped barium titanate ($BaTiO_3$) nanoparticles. *Mater. Today Proc.*, S2214785323052379. doi: 10.1016/j.matpr.2023.11.131.
26. Lin, Y., et al., (2019). Excellent energy-storage properties achieved in $BaTiO_3$-based lead-free relaxor ferroelectric ceramics via domain engineering on the nanoscale. *ACS Appl. Mater. Interfaces*, *11*. doi: 10.1021/acsami.9b10819.
27. Emelianov, N. A., Korotkov, L. N., Al Mandalawi, W. M., Kashirin, M. A., Roldan, L. J. A., & Korotkova, T. N., (2018). Ferromagnetism in barium titanate nanoparticles. Effect of annealing in hydrogen atmosphere on magnetic properties. In: *2018 IEEE ISAF-FMA-AMF-AMEC-PFM Joint Conference (IFAAP)* (pp. 1–3). Hiroshima: IEEE. doi: 10.1109/ISAF.2018.8463349.
28. Jiang, B., et al., (2019). Barium titanate at the nanoscale: Controlled synthesis and dielectric and ferroelectric properties. *Chem. Soc. Rev.*, *48*(4), 1194–1228. doi: 10.1039/C8CS00583D.
29. Karvounis, A., Timpu, F., Vogler-Neuling, V. V., Savo, R., & Grange, R., (2020). Barium titanate nanostructures and thin films for photonics. *Adv. Opt. Mater.*, *8*(24), 2001249. doi: 10.1002/adom.202001249.
30. Korkmaz, S., & Kariper, I. A., (2022). $BaTiO_3$-based nanogenerators: Fundamentals and current status. *Electroceramics J.*, *48*(1), 8–34. doi: 10.1007/s10832-021-00266-3.
31. Jayakrishnan, A. R., et al., (2023). Are lead-free relaxor ferroelectric materials the most promising candidates for energy storage capacitors? *Prog. Mater. Sci.*, *132*, 101046. doi: 10.1016/j.pmatsci.2022.101046.
32. Saressalo, A., (2016). *Ferroelectric Properties of BFBT(Mn) Thin Films*.
33. Liu, T., et al., (2022). Excellent energy storage performance in $Bi(Fe_{0.93}Mn_{0.05}Ti_{0.02})O_3$ modified $CaBi_4Ti_4O_{15}$ thin film by adjusting annealing temperature. *Nanomaterials*, *12*(5), 730. doi: 10.3390/nano12050730.
34. Ishiwara, H., (2012). Ferroelectric random access memories. *J. Nanosci. Nanotechnol.*, *12*, 7619–7627. doi: 10.1166/jnn.2012.6651.
35. *Recent Advances in Multilayer-Structure Dielectrics for Energy Storage Application –* PMC. [Online]. Available: https://www.ncbi.nlm.nih.gov/pmc/articles/PMC8655226/ (accessed on 24 February 2024).
36. Liu, M., Zhu, H., Zhang, Y., Xue, C., & Ouyang, J., (2016). Energy storage characteristics of $BiFeO_3/BaTiO_3$ Bi-layers integrated on Si. *Materials*, *9*(11), 935. doi: 10.3390/ma9110935.
37. Yuan, M., Zhang, W., Wang, X., Pan, W., Wang, L., & Ouyang, J., (2013). In situ preparation of high dielectric constant, low-loss ferroelectric $BaTiO_3$ FILMS on Si at 500°C. *Appl. Surf. Sci.*, *270*, 319–323. doi: 10.1016/j.apsusc.2013.01.023.

38. Zhu, H., et al., (2016). Microstructural and electrical characteristics of epitaxial $BiFeO_3$ thick films sputtered at different Ar/O_2 flow ratios. *CrystEngComm*, *18*(24), 4604–4612. doi: 10.1039/C6CE00781C.
39. Wang, Y., et al., (2020). Design strategy of barium titanate/polyvinylidene fluoride-based nanocomposite films for high energy storage. *J. Mater. Chem. A*, *8*(3), 884–917. doi: 10.1039/C9TA11527G.
40. Sharma, S., Tomar, M., Puri, N. K., & Gupta, V., (2016). $BiFeO_3$ /$BaTiO_3$ multilayer structures for solar energy harvesting application. *Energy Harvest. Syst.*, *3*(3), 237–243. doi: 10.1515/ehs-2016-0001.
41. Toupet, H., Shvartsman, V., Le Marrec, F., Borisov, P., Kleemann, W., & Karkut, M., (2008). Enhanced magnetization in $BiFeO_3$/$BaTiO_3$ multilayers: An interface effect? *Integr. Ferroelectr.*, *100*, 165–176. doi: 10.1080/10584580802541023.

CHAPTER 4

Nanostructured Electrocatalysts for Green Hydrogen

K. NAYANA[1,2] and A. P. SUNITHA[1]

[1]*Department of Physics, Government Victoria College, Palakkad, Kerala, Affiliated to the University of Calicut, India*

[2]*Department of Physics, N. S. S. College, Ottapalam, Palakkad, Kerala, Affiliated to the University of Calicut, India*

ABSTRACT

Hydrogen is a non-polluting, non-conventional alternative in the fuel industry. The entire world is conducting research on the effective production, storage, and application of Hydrogen to meet the energy needs of society. Numerous hydrogen policies have been implemented by various countries. In this context, the production of Hydrogen through green methods is a significant requirement. This chapter will discuss the research developments on water splitting using nanostructured dichalcogenides materials as electrocatalysts.

4.1 INTRODUCTION

The International Energy Agency has pointed out that the world currently relies on fossil fuels for 80% of its energy production. This results in the release of 30–35 Gt of CO_2 annually [1]. However, conventional methods of energy production and distribution face numerous challenges such as limited availability of raw materials, technological limitations, inefficiency

Sustainability in Energy and Environment: Engineered Materials and Smart Computational Techniques. Ann Rose Abraham, Heru Susanto, A. K. Haghi, & Kian Hariri Asli (Eds.)

in production methods, pollution from production processes, and cost-effectiveness [2–5]. In the event of natural disasters or disruptions to conventional electricity supply in a region, the process of restoring electricity can be time-consuming and expensive [6]. Additionally, rural areas often lack sufficient energy production methods, leading to increased hardships for the local population [7, 8]. Therefore, it is crucial to develop non-conventional, renewable, cost-effective, and eco-friendly energy production methods to meet the future energy needs of the world [2–4].

The alternative methods of energy production mainly rely on non-conventional sources such as solar cells, wind energy, and more. The low efficiency and high cost of solar cells and windmills necessitate the exploration of other power sources. One emerging alternative method for energy production is hydrogen fuel cells [2, 5]. Extensive research is being conducted worldwide to study hydrogen production, storage, and utilization. The development of hydrogen power plants and hydrogen trade is currently on the rise. China is currently constructing the world's largest green hydrogen power plant, with a cost of 2.6 USD per kilogram of hydrogen. The plant is expected to be operational by June 2023 [9].

Different types of hydrogen, electrochemical mechanisms involved in the production of hydrogen, the working of a hydrogen fuel cell, and a briefing of hydrogen policies of nations are explained in this chapter. Efficient hydrogen production requires the presence of a catalyst to enhance its production rate. Nano-electrocatalysts and their activity parameters in hydrogen production mechanisms are discussed here.

4.2 WHY HYDROGEN?

Hydrogen is the most abundant element in nature. However, the existence of pure hydrogen molecules in nature is only about 0.00005% [10, 11]. It exists in other forms such as water and hydrocarbons. Therefore, we must extract hydrogen from compounds that contain it. The process of hydrogen molecules evolving from their compounds is known as the hydrogen evolution reaction (HER) [12, 13]. In a hydrogen fuel cell, the chemical energy stored in the bonds of hydrogen molecules is utilized [12–14]. The fuel value of hydrogen is 141.9 KJ/g at 1 atm pressure and 25°C temperature [15–17]. Hydrogen has a high power density of 2 KWh/Kg [18]. It is possible to produce hydrogen in an eco-friendly manner with a very low carbon footprint.

4.3 DIFFERENT TYPES OF HYDROGEN

The quality of hydrogen is labeled based on the production methods. The methods to produce hydrogen include steam reforming of natural gases, coal gasification, and water electrolysis [14, 19, 20]. Over 90% of hydrogen production comes from coal gasification and steam reforming [12]. Hydrogen produced by steam reforming is known as grey hydrogen [19]. Hydrogen produced from coal is known as brown hydrogen. Both grey and brown hydrogen production result in a large quantity of carbon and greenhouse gases (GHGs). This type of hydrogen can be converted to blue hydrogen by using carbon capture techniques [19]. A hydrogen production technique with nearly zero carbon footprint is known as green hydrogen [3, 21, 22]. Since water is the cheapest and most abundant compound containing hydrogen, water splitting is the simplest process for producing hydrogen [22]. Water electrolysis, assisted by non-conventional energy sources, can produce green hydrogen [2]. However, the quality of synthesized hydrogen remains a significant challenge. Researchers worldwide are working diligently on the synthesis of green hydrogen.

Non-conventional energy sources such as solar cells and windmills can supply energy to the HER reaction. Minimizing energy consumption is one of the important parameters discussed for cost-effective hydrogen production. Catalysts are used to enhance the reaction rate [23, 24]. Since the carbon footprint of solar cells and wind energy is considered zero, the corresponding hydrogen produced is purely green [25–27].

4.4 APPLICATION OF HYDROGEN IN ENERGY PRODUCTION

4.4.1 HYDROGEN FUEL CELL

Hydrogen fuel cell is a two-electrode electrochemical cell. The hydrogen molecule injected into the fuel cell will split into a hydrogen ion and an electron near the anode. At the cathode, the hydrogen ion and electron can combine with atmospheric oxygen to form water. During these reactions, current will flow from the cathode to the anode. Water and heat energy are the by-products of the fuel cell. This is the basic working principle of a hydrogen fuel cell [28, 29]. However, for an unlimited supply of pure green hydrogen to the fuel cell, the use of the best catalyst to enhance the reaction rate is very important [24, 25]. The reaction should be eco-friendly and economically feasible. These factors make the design of the fuel cell more complicated [29]. Figure 4.1(a) and 1(b) represent the working of the catalyst-assisted water splitting electrolysis system and the hydrogen fuel cell, respectively [29].

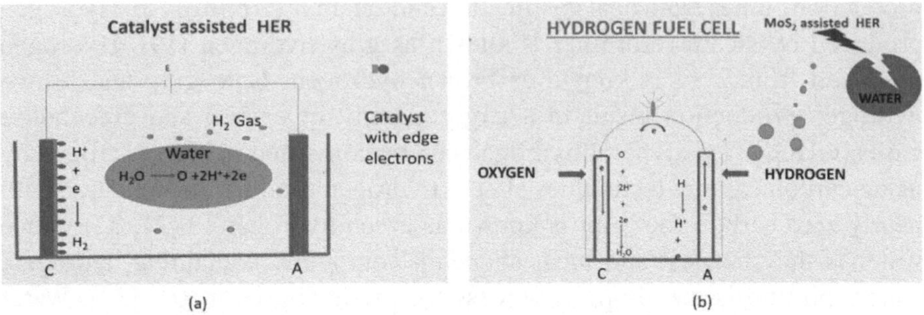

FIGURE 4.1 (a) Electrocatalyst-assisted water electrolysis; and (b) operation of a hydrogen fuel cell.

The carbon footprint of a hydrogen fuel cell can be reduced to zero if it is powered by non-conventional energy sources [30, 31]. It is feasible to develop compact hydrogen fuel cells to provide power to rural areas and replace conventional electricity [7, 8]. However, the challenges in hydrogen production and storage limitations make fuel cells more expensive [3–5]. To develop cost-effective and user-friendly hydrogen fuel cells, it is necessary to overcome these issues with hydrogen production and storage [3–5].

4.4.2 HYDROGEN FUEL CELL POWERED VEHICLES

Recently, companies such as Mercedes, BMW, Ford, and Toyota have developed hydrogen fuel cell powered cars [32–34]. These vehicles require tanks to store hydrogen. There are various types of tanks available, including high pressure hydrogen tanks, liquefied hydrogen tanks, hydrogen absorbing alloy tanks, and hybrid tanks [32]. In high pressure hydrogen tanks, hydrogen is stored in compressed gas form [32, 35]. The advantages of high-pressure hydrogen tanks are their simple structure and charge-discharge process for hydrogen [35]. However, these tanks can be expensive [35]. Hydrogen powered vehicles also use liquid hydrogen tanks to store high density liquid hydrogen [32, 35]. A tank with a storage capacity of 70 MPa can allow for a travel distance of approximately 50 km [32]. Hydrogen-absorbing alloy tanks can store hydrogen at a higher density than liquid hydrogen [32, 35]. A vehicle powered by a hydrogen fuel cell requires a supply of approximately 1.1 g of pure hydrogen per second [36]. Hydrogen fuel cell-powered vehicles have achieved an efficiency ranging from 30% to 45% [35]. According to a joint agency report by the California Energy Commission, 1 kg of hydrogen has a price of $16 in California [29].

4.5 HYDROGEN POLICIES

In 2020, the Hydrogen Council held in Belgium decided to achieve 18% of total energy consumption through hydrogen fuel by 2050. In the same year, 18 countries published a hydrogen roadmap [37]. The frequent catastrophic events in Japan prompted them to establish a hydrogen society as a solution to overcome electricity failures. According to Japan's roadmap for hydrogen fuel, they are focusing on the development of household fuel cell appliances and fuel cell vehicles [38]. To accelerate the development of hydrogen fuel stations and hydrogen fuel cell vehicles, Japan has initiated a collaboration among various infrastructure development companies [38]. The Global Hydrogen Review (2022), published by the International Energy Agency, provides insights into the current hydrogen policies and regulations implemented by different governments to promote the use of hydrogen as a fuel source [1]. Efforts are underway to develop infrastructure for hydrogen fuel handling, establish an international trade market, define standards, certificates, and rules, and address regulations. Clearing the uncertainty surrounding world trade organization trade rules for hydrogen is crucial for establishing international hydrogen exports. International cooperation for the establishment of hydrogen trade is rapidly growing [1].

4.6 NANO-ELECTROCATALYSTS FOR HYDROGEN EVOLUTION

In order to break a water molecule with electrical energy, a minimum energy of 1.23 eV needs to be supplied. This energy is associated with the Gibb's free energy of water splitting [2, 23, 39, 40]. However, in practical scenarios, the energy required will be higher than 1.23 eV due to solution resistance and electrode resistance [2]. The additional potential applied for water splitting is referred to as overpotential. The presence of a catalyst helps to reduce the value of overpotential.

Using the advantage of nanotechnology, new energy materials can be fabricated to enhance the reaction kinetics of water splitting, reducing the energy supplied to the electrolysis system. These nanomaterials (NMs) are known as nano-catalysts. Due to their high active surface area compared to their bulk, nano-catalysts exhibit increased catalytic reaction rates [18, 41]. Different types of catalysts exist depending on the mechanism they offer. Electrocatalysts are catalysts that intensify the reaction kinetics of HER with the help of an external biasing voltage [42, 43]. Some catalysts can enhance the HER mechanism by absorbing energy from sunlight, known as photocatalysts [16, 45]. Certain NMs can utilize both electrocatalytic and photocatalytic properties simultaneously, making them photoelectrocatalysts [46–49].

There were many NMs synthesized to assess their electrocatalytic properties. Among them, various compounds of transition metals (TMs) have been found to be effective for electrocatalysis. Oxides, sulfides, phosphides, nitrides, selenides, and carbides of TMs such as iron, copper, nickel, cobalt, molybdenum, and tungsten are widely used as electrocatalysts. Researchers have developed nanostructured transition metal compounds with different structures, morphologies, and compositions. The catalytic activity of these catalysts is determined by engineering the number of active edge sites on the catalyst surface and the catalyst's conductivity. Catalysts with a high density of active sites include those that are porous, defect-rich, or have deficiency sites [2]. Doping and heterostructures are examples of catalyst modifications that enhance conductivity. Some catalysts exhibit different catalytic activities depending on their phase.

4.6.1 HYDROGEN EVOLUTION REACTION (HER) MECHANISM

The Volmer reaction, Heyrovsky reaction, and Tafel reaction are the three possible reaction mechanisms involved in HER. The HER mechanism for water splitting depends on the pH of water. In acidic conditions, the adsorption of hydrogen ions on the surface of the catalyst is known as the Volmer reaction or discharge reaction (Eqn. (1)) [50, 51].

$$H^{+} + e^{-} = H_{ads} \tag{1}$$

For the fast Volmer reaction, the catalyst should contain a large number of active edge sites to adsorb hydrogen ions. The presence of plenty of active edge sites in a nano-electrocatalyst can enhance the Volmer reaction [52]. After the Volmer reaction, the adsorbed hydrogen ion can either undergo the

Heyrovsky reaction or the Tafel reaction [53]. In the Heyrovsky reaction, the adsorbed hydrogen ion combines with an electron and another hydrogen ion to form a hydrogen molecule, as shown below:

$$H_{ads} + e^- + H^+ = H_2 \tag{2}$$

In Tafel reaction, two adsorbed hydrogen ions combine to form hydrogen molecule (Eqn. (3)).

$$2H_{ads} = H_2 \tag{3}$$

In alkaline or neutral water, there are no hydrogen ions. Therefore, the splitting of water molecules initiates the HER mechanism. The Volmer, Heyrovsky, and Tafel reaction mechanisms for alkaline media are given in Equations (4)–(6), respectively.

$$H_2O + e^- = H_{ads} + OH^- \tag{4}$$

$$H_{ads} + H_2O + e^- = OH^- + H_2 \tag{5}$$

$$2H_2O + 2e^- = 2OH^- + H_2 \tag{6}$$

where; H_{ads} represents adsorbed hydrogen ion on catalyst surface. HER either follow Volmer-Heyrovsky mechanism or Volmer-Tafel mechanism.

In acidic media, the rate of the Volmer reaction depends solely on the reduction potential of the hydrogen ion. However, in non-acidic media, the Volmer reaction rate is influenced by the potential for water dissociation, reduction potential of the hydrogen ion, and desorption potential of the hydroxide ion. Therefore, in most cases, an acidic electrolyte is preferred for water splitting. Additionally, an acidic electrolyte facilitates good conductivity for ion passage [54–56].

4.6.2 PROPERTIES OF A GOOD CATALYST USED FOR SPLITTING WATER

Gibbs' free energy of adsorption of hydrogen on the surface of a catalyst should be very low for it to be considered a good catalyst. Theoretically, the best catalyst would have a zero Gibbs' free energy of adsorption [57]. Platinum is the best electrocatalyst for HER because the Gibbs' free energy of adsorption of hydrogen on platinum is only about –0.09 eV [57]. However, platinum is very expensive. Therefore, it is highly desirable to produce economically feasible, naturally abundant, and eco-friendly non-metallic materials as electrocatalysts. Materials with a strong negative Gibbs' free energy easily undergo adsorption of hydrogen ions on the surface (Volmer reaction). However, due to the strong attraction of Hads on the catalytic site, the rate of hydrogen desorption reaction (either Heyrovsky or Tafel) decreases. Materials with a strong positive Gibbs' energy have a weak adsorption rate of hydrogen ions, which negatively affects the overall reaction rate. Therefore, the Gibbs' free energy of the material should be very small for it to be the best electrocatalyst. In such a catalyst, Hads is not strongly or weakly bonded at the catalyst site [56, 57].

According to density functional theory, the expression for Gibbs' free energy of adsorption or desorption ΔG is written as [58]:

$$\Delta G = \Delta E^1 + \Delta E^2 - T\Delta S \tag{7}$$

where; ΔE^1 is the energy between catalyst; and H_{ads}. ΔE^2 is the zero-point energy between H_{ads} and isolated hydrogen ion; 'T' is the temperature; and ΔS is the entropy change during adsorption or desorption. Noskov et al. determined the ΔG value of platinum and some other metals and designed a volcano plot between ΔG and exchange current density [59, 60].

4.6.3 HYDROGEN EVOLUTION REACTION (HER) ACTIVITY PARAMETERS

There are so many activity parameters to determine the kinetics of the electrochemical reaction of the HER mechanism.

4.6.3.1 OVERPOTENTIAL

Overpotential is the excess potential compared to the dissociation energy of water (1.23 eV) that needs to be applied to produce hydrogen molecules. This excess potential arises from the solution resistance and electrode resistances of the electrolysis cell. The minimum value of overpotential indicates a faster HER. Overpotential is measured at a specific current density, representing the amount of potential that needs to be excessively applied to generate that current density. Typically, overpotential at 10 mA/cm^2, 20 mA/cm^2, or 100 mA/cm^2 is reported [61–64]. Onset overpotential is another parameter commonly used by researchers, indicating the overpotential at a current density of 0.5 mA/cm^2 or 1 mA/cm^2 [23, 65–67]. The current density at zero overpotential is known as the exchange current density [67, 68]. Materials with minimum Gibbs' free energy have the maximum exchange current density [58].

4.6.3.2 TAFEL SLOPE

The plot of the logarithm of current density versus overpotential is known as the Tafel plot. The linear portion of the Tafel plot can be mathematically expressed as:

$$Over\ potential = a + b * \log(J) \tag{8}$$

where 'a' is known as the Tafel slope, which indicates the rate of applied potential to increase the current density 'J.' A low value of the Tafel slope means better kinetics of HER [58, 66]. The value of the Tafel slope gives an idea about the rate-determining reaction mechanism involved in HER. The reaction with the lowest rate will determine the overall rate of HER. Therefore, the reaction with the lowest rate is known as the rate-determining step (RDS). The HER reaction mechanism may follow four reaction paths. Volmer-Heyrovsky or Volmer-Tafel reactions with either Volmer, Heyrovsky, or Tafel as RDS [53, 56] are among them. For Volmer and Heyrovsky as RDS, the reaction kinetics follow the Butler-Volmer equation (Eqn. (9)):

$$i = i_0 \left(-e^{\frac{\alpha z F \eta}{RT}} + e^{\frac{\beta z F \eta}{RT}} \right) \tag{9}$$

where 'i' is the current density through the electrode, 'i0' is the exchange current density, 'α' and 'β' are the cathode and anode charge transfer coefficients, η is the overpotential, 'R' is the universal gas constant, 'F' is the Faraday's constant, 'z' is the number of electrons participating in the reaction, and 'T' is taken as the temperature (298 K). Generally, the values of α and β are taken as 0.5 [53, 56, 68]. For the Tafel reaction, there is no charge transfer occurring. Therefore, the Butler-Volmer equation has no significance in the Tafel reaction as the rate-determining step. The first and second exponential terms in Eqn. (9) represent the forward and backward reactions. For fast HER, the rate of the backward reaction is very low compared to the rate of the forward reaction [56].

For Volmer-Heyrovsky reaction mechanism with Volmer as the RDS, Butler-Volmer equation reduces to:

$$i = 2i_0 \left(-e^{\frac{\alpha F \eta}{RT}} + e^{\frac{(1+\beta) F \eta}{RT}} \right) \tag{10}$$

For the same Volmer-Heyrovsky reaction with Heyrovsky as the rate-determining step (RDS), the Butler-Volmer equation is reduced to:

$$i = 2i_0\left(-e^{\frac{(1+\alpha)F\eta}{RT}} + e^{\frac{\beta F\eta}{RT}}\right) \tag{11}$$

In the Volmer-Tafel reaction mechanism with Volmer as the rate-determining step (RDS), the Butler-Volmer equation is reduced to:

$$i = 2i_0\left(-e^{\frac{-\alpha F\eta}{RT}} + e^{\frac{\beta F\eta}{RT}}\right) \tag{12}$$

Tafel slope derived from the Eqns. (10) and (12) are:

$$\frac{-2.303RT}{\alpha F} \tag{13}$$

Where, Volmer is the RDS. Tafel slope for Eqn. (11) is:

$$\frac{-2.303RT}{(1+\alpha)F} \tag{14}$$

where; Heyrovsky is the RDS. The corresponding Tafel slopes of Eqns. (13) and (14) at 298 K are 120 mV/Dec, 40 mV/Dec, respectively. For the Volmer-Tafel reaction with Tafel as RDS, Tafel slope will be:

$$\frac{-2.303RT}{2F} \tag{15}$$

The value of the Tafel slope is 30 mV/Dec. Therefore, if the Tafel plot of the catalyst is above 120 mV/Dec, the rate-determining step (RDS) will be the Volmer reaction. In the case of the hydrogen evolution reaction (HER), the desorption of Hads will occur faster than adsorption. If the Tafel slope is between 40 mV/Dec. and 120 mV/Dec., the Heyrovsky mechanism will be the RDS, where adsorption is faster than desorption. If the Tafel slope is below 40 mV/Dec, the RDS is the Tafel step with faster adsorption than the formation of hydrogen molecules. For a platinum catalyst, the Tafel slope is approximately 30 mV/Dec. [53, 56, 69].

4.6.3.3 TURN OVER FREQUENCY (TOF)

Turnover frequency (TOF) is the amount of hydrogen molecules evolved per second per catalytic site. Calculating an accurate value of TOF is not easy, as we cannot predict the total number of active sites of the catalyst participating in the HER. However, we assume that all active edges of the catalyst take part in the catalytic activity. This assumption may lead to an inaccurate result in the TOF value. Nonetheless, it is easy to compare two or more catalysts in terms of the estimated TOF value [2]. The approximate value of TOF can be calculated using Equation (16) [23]:

$$TOF = \frac{I}{2Fn} \tag{16}$$

where 'I' is the current, 'F' is Faraday's constant, and 'n' is the number of moles of catalyst coating on the electrode. The dividing factor of '2' indicates that two Hads are necessary for the evolution of one hydrogen molecule.

To find out overpotential, Tafel slope and TOF, linear sweep voltammetry is used.

4.6.3.4 DURABILITY OF REACTION

The durability of a reaction or the constancy in reaction rate is an important factor in HER. It can be represented either in terms of the duration of the un-decayed reaction rate or in terms of the number of cycles in cyclic voltammetry (CV) with the same area. Chronopotentiometry/chronoamperometry measurements give the lifespan of the un-decayed reaction rate [70–72]. Continuous measurement in CV for thousands of cycles, up to which its area starts to decay, gives the durability of the reaction rate [70, 73]. A constant water splitting rate will result in a uniform emission of hydrogen.

4.6.3.5 DOUBLE LAYER CAPACITANCE

In a water electrolysis system, the electrode-electrolyte interface provides a capacitance known as the Stern layer capacitance [74]. The remaining electrolyte, which contains diffusing ions, offers another capacitance known as the diffuse layer capacitance. The combined resistance of the Stern layer capacitance and the diffuse layer capacitance is referred to as the double layer capacitance. The electrode material and the ions of the electrolyte attracted to the surface of the electrode act as the two conducting plates of the Stern capacitance. Solvent molecules serve as the dielectric medium of the Stern capacitance. Since the conducting plates and the dielectric of Stern capacitor are static, stern capacitance is electrostatic capacitance. But diffusing ions can move through the electrolyte. So diffused layer capacitance is dynamic capacitance [74–77]. The concentration of electrolyte inversely affects the contribution of the diffuse layer capacitance to the double layer capacitance [76].

Double layer capacitance depends on the number of charges actively participating in electrolysis. Therefore, a higher value of capacitance indicates a higher activity of HER. Double layer capacitance can be calculated using either Equation (17) [78] or Equation (18) [61, 76].

$$C = \frac{A}{(SR)(dV)} \tag{17}$$

$$C = \frac{Q}{dV} \tag{18}$$

where 'A' represents the area of the curve plotted between current density and potential, 'SR' denotes the scan rate of CV, which indicates the rate of potential variation applied with respect to time. 'V' represents the potential window applied in CV, while 'Q' represents the actively participating charge in the electrode. The area of the plot between current and time in CV provides the charge 'Q.' A plot of scan rate versus current density at an overpotential can be generated using CV. The slope of the same curve will yield the double layer capacitance. In the case of the existence of double-layer capacitance, no ion exchange occurs between the electrode and electrolyte. Therefore, double-layer capacitance results in capacitive current. If there is an ion exchange between the electrode and electrolyte, oxidation-reduction peaks will appear in the cyclic voltammograms. This leads to pseudo capacitance in the electrolysis cell, resulting in faradaic current [78–80].

4.6.3.6 CHARGE TRANSFER RESISTANCE

Charge transfer resistance is the resistance offered by the electrode-electrolyte interface in an electrolysis cell. Lowering the value of charge transfer resistance boosts the HER rate. Charge transfer resistance is calculated from the Nyquist/Bode plot using electrochemical impedance spectroscopy (EIS). The low-frequency end of the Nyquist/Bode plot gives the value of the sum of charge transfer resistance and solution resistance. Solution resistance is the resistance offered by the electrolyte. The high-frequency end of the Nyquist/Bode plot gives the value of the solution resistance [41, 47, 82].

4.7 APPLICATION OF NANOSTRUCTURED TRANSITION METAL COMPOUNDS ELECTROCATALYST FOR HYDROGEN EVOLUTION

4.7.1 TRANSITION METAL SULFIDES

Transition metal sulfides have a layered structure. The covalent bond between the metal and sulfur is strong enough to form stable molecules. However, each layer of transition metal sulfide molecules is interconnected by Van der Waal's forces. By breaking these weak Van der Waal's forces, it is possible to create nanostructures from bulk materials [82-84]. The bottom-up approach is commonly used to synthesize nanostructures of transition metal sulfides, resulting in catalysts with various morphologies.

Iron, copper, nickel, cobalt, tungsten, and molybdenum sulfides are used as electrocatalysts for HER. However, among these, molybdenum sulfide is a very promising electrocatalyst with high activity. The ease of hydrogen ion adsorption on the active sites of MoS_2 nanostructure makes it a promising catalyst. The lone pair of electrons in the sulfur atom acts as the active sites of MoS_2. There are three phases of MoS_2 that exist ($1TMoS_2$, $2HMoS_2$, and $3RMoS_2$). Here, $1TMoS_2$ exhibits metallic behavior and is also unstable [82]. On the other hand, $2HMoS_2$ and $3RMoS_2$ exhibit semiconducting behavior and are stable. $3RMoS_2$ is HER inactive, while the nanostructure of $2HMoS_2$ is HER active. However, only the active edge sites of $2HMoS_2$ participate in HER activity, as the basal planes of $2HMoS_2$ are HER inactive. The basal plane of metallic $1TMoS_2$ is HER active. Through lithium intercalation or precise synthesis methods, $1TMoS_2$ can be constructed [85]. Extensive research has demonstrated the efficiency of MoS_2 nanostructures in HER. Heterostructures of MoS_2 with graphene, reduced graphene oxide, and tungsten disulfide also exhibit high electrocatalytic HER activity. Due to tungsten and molybdenum belonging to the same group (Group 6) in the periodic table, tungsten also serves as a highly efficient electrocatalyst for HER, owing to similarities in electronic structure. Iron, copper, cobalt, and nickel sulfides have also been extensively studied for the same purpose. Table 4.1 provides examples of transition metal sulfides with good electrocatalytic activity.

4.7.2 TRANSITION METAL SELENIDES

Sulfur and selenium belong to group VI A elements in the periodic table. Their similar electronic configuration in the outer electronic state results in

TABLE 4.1 Electrocatalytic Activity of Transition Metal Disulfides

Electrocatalyst	Overpotential at Current Density (mV)@mA/cm^2	Tafel Slope (mV/Dec)	TOF at Overpotential (s^{-1}) @mV	Durability (Cycles) or Hours	Capacitance (mF/cm^2)	Charge Transfer Resistance (Ω)	References
Pore rich monolayer MoS_2	241 @10	163	/	2,000 s	9.89	3.8 K	[61]
Defective MoS_2QDs	187 @100	53	/	16,000 cycles	/	5	[83]
Amorphous MoS_2QDs	300 @0.1	73.9	2.83	12 hr.	/	64	[84]
Mn doped MoS_2/rGO	20 @4.52	76	0.042	1,000 cycles	17.37	1.5	[86]
MoO_3-MoS_2nanowire	150 @1	40–50	4	10,000 cycles	/	14	[87]
Defect rich of MoS_2	71.5 @10	58.3	/	2,000 cycles	4.7	26.1	[88]

similar chemical behavior. Selenium has a lower ionization energy and better conductivity compared to sulfur. As a result, compounds like selenides of molybdenum, tungsten, iron, cobalt, and nickel have received significant attention in electrocatalytic activity [2, 89]. Table 4.2 provides the activity table of some reported transition metal selenides.

4.7.3 TRANSITION METAL CARBIDES

When carbon is bonded with transition metals to form transition metal carbides, its electronic configuration favors the HER activity. The presence of carbon in transition metal compounds creates more active edge sites, enhancing conductivity. As a result, transition metal carbides are popular as HER electrocatalysts [2, 68]. Transition metal carbides exhibit HER activity across a wide pH range. Platinum, which is the best electrocatalyst for HER in acidic media, is not as effective in alkaline media. However, transition metal carbides demonstrate highly intensive electrocatalytic activity in neutral and alkaline media. Tungsten carbides and molybdenum carbides have been successfully fabricated by numerous researchers, exhibiting good catalytic properties. Table 4.3 provides a list of the electrocatalytic activity of some transition metal carbides.

4.7.4 TRANSITION METAL PHOSPHIDES

The structure of transition metal (TM) phosphides is trigonal prismatic [2], meaning they do not have a layered structure. Unlike layered structures found in transition metal dichalcogenides (TMDCs), the basal planes of TM phosphides are HER inactive. This structural difference contributes to their high electrocatalytic activity. TM phosphides also exhibit high electrical conductivity, metallic behavior, and stability [54, 58]. Table 4.4 provides information on the electrocatalytic activity of transition metal phosphides.

4.7.5 TRANSITION METAL NITRIDES

Transition metal nitrides and their heterostructures are beneficial for the hydrogen evolution reaction (HER). Molybdenum, tungsten, nickel, and

TABLE 4.2 Electrocatalytic Activity of Transition Metal Diselinides

Electrocatalyst	Overpotential at Current Density (mV)@mA/cm^2	Tafel Slope (mV/Dec)	TOF at Overpotential (s^{-1}) @mV	Durability (Cycles) or Hours	Capacitance (mF/cm^2)	Charge Transfer Resistance (Ω)	References
$Ni_{0.5}Mo_{0.5}Se$	190 @10	107	/	1,000 cycles	1.21	86.5	[90]
$FeSe_2$-$MoSe_2$/rGo	101 @10	55	/	1,000 cycles	8.4	3.2	[91]
$Mo_3Se_{12}S$	253 @10	84	/	1,000 cycles	25.7	/	[92]
$MoCoSe_x$	60 @100	67	/	1,000 cycles	6	/	[93]
$Ni_{0.33}Co_{0.67}Se_2$	65 @10	35	/	24 hr.	197	0.27	[94]
Co-$FeSe_2$	129 @10	47	/	2,000 cycles	35.4	/	[95]

TABLE 4.3 Electrocatalytic Activity of Transition Metal Carbides

Electrocatalyst	Overpotential at Current Density (mV)@mA/cm^2	Tafel Slope (mV/Dec)	TOF at Overpotential (s^{-1}) @mV	Durability (Cycles) or Hours	Capacitance (mF/cm^2)	Charge Transfer Resistance (Ω)	References
Mo_2C	138 @10	85	/	5,000 cycles	4	/	[97]
CoMoC	46 @10	46	/	10,000 cycles	/	/	[98]
γ–MoC	/	121.6	/	1,000 cycles	/	/	[99]
Tungsten oxide/carbide	233 @20	69	/	6 hr.	/	/	[100]
Mo_2C at nitrogen doped graphene.	135 @10	65	/	1,000 cycles	8.1	/	[101]

TABLE 4.4 Electrocatalytic Activity f Transition Metal Phosphides

Electrocatalyst	**Overpotential at Current Density (mV)@mA/cm^2**	**Tafel Slope (mV/Dec)**	**TOF at Overpotential (s^{-1}) @mV**	**Durability (Cycles) or Hours**	**Capacitance (mF/cm^2)**	**Charge Transfer Resistance (Ω)**	**References**
Oxygen incorporatedCo_2P	160 @10	61.1	/	1,000 cycles	/	/	[103]
Nickel phosphide nanosheet.	10 @10	79.1	/	1,000 cycles	/	289.9	[104]
Ni_2P	130 @20	46	0.5 @200	500 cycles	/	/	[105]
Mn doped CoP	65 @10	46.16	/	24 hr.	293.75	/	[106]
MoP	/	54	/	40 hr.	/	/	[107]
Fe_2P	147 @10	62	/	6,000 s	/	3.18	[108]

copper-based nitrides have been synthesized and electrochemically tested for HER. Nickel-molybdenum-nitride, synthesized by Wu et al. [22], and Chang et al. [109], showed activity for overall water splitting. Defect-rich molybdenum nitride, synthesized by Xiong et al. [110], exhibited good electrocatalytic properties to enhance HER. The heterostructure of reduced graphene and phosphorus-modified tungsten nitride, produced by Yan et al. [111], and the copper-nickel-nitride compound synthesized by Wang et al. [112], also indicate that different transition metal nitrides have unique HER boosting properties. However, determining the position and quantity of active edge sites in nitrides remains challenging. Therefore, further research efforts are needed to obtain a clear understanding of the catalytic activity of transition metal nitrides for HER. Table 4.5 provides a summary of reported works on transition metal nitrides as electrocatalysts for HER.

4.7.6 TRANSITION METAL OXIDES

Transition metal oxides are emerging as electrocatalysts for the hydrogen evolution reaction (HER). Doping, morphology engineering, and oxygen deficiency control are the main strategies adopted to control the parameters of HER activity [113]. By inducing oxygen deficiency, the electronic conductivity and hydrogen adsorption energies of the catalyst can be tuned [114]. Doping can also alter the number of catalytic activity sites to enhance the HER [115]. However, metal oxides are unstable in acidic electrolytes, so most studies focus on alkaline electrolytes [23]. Table 4.6 provides the activity data for transition metal oxides.

4.7.7 TRANSITION METAL DICHALCOGENIDE/CARBON DERIVATIVE HETEROSTRUCTURES

Nanocomposites of carbon are popular for their use as HER electrocatalysts. Heterostructures of carbon nanotubes (CNTs), graphene, graphene oxides, or reduced graphene oxide can enhance the performance of the electrocatalyst. The introduction of carbon derivatives to the catalyst can enhance electric conductivity. Graphene or reduced graphene oxide layers create a conductive path for electrons to participate in catalytic activity [118]. The strong bonding between carbon and metal atoms in the electrocatalyst allows for

TABLE 4.5 Electrocatalytic Activity of Transition Metal Nitrides

Electrocatalyst	Overpotential at Current Density (mV)@mA/cm^2	Tafel Slope (mV/Dec)	TOF at Overpotential (s^{-1}) @mV	Durability (Cycles) or Hours	Capacitance (mF/cm^2)	Charge Transfer Resistance (Ω)	References
Ni_3N-NiMoN	31 @10	64	0.265 @100	5,000	241	/	[22]
NiMoN nanowire	47 @10	54	/	2,000	82.68	/	[109]
Defect rich MoN	125 @10	51.15	/	2,000	18.59	/	[110]
P-WN/rGO	265 @10	54	/	5,000	32	/	[111]
Copper-nickel-nitride	71.4 @10	106.5	0.49 @150	/	42.02	/	[112]

TABLE 4.6 Electrocatalytic Activity Parameters of Transition Metal Oxides

Electrocatalyst	Overpotential at Current Density (mV)@mA/cm^2	Tafel Slope (mV/Dec)	TOF at Overpotential (s^{-1}) @mV	Durability (Cycles) or Hours	Capacitance (mF/cm^2)	Charge Transfer Resistance (Ω)	References
$WO_{2.9}$	70 @10	50	8.04 @100	1,000 cycles	/	/	[114]
Nitrogen doped NiO	100 @10	112.58	/	2,000 cycles	/	12.8	[115]
$TiO_2/Co_2P_4O_{12}$	81 @20	87	1 @52	40 hr.	55.1	11.1	[25]
Ag/Cu_2O	108	58	/	20 hr.	44.9	/	[117]

good dispersion in the carbon composite, increasing the number of active edges of the catalyst [23, 118]. Additionally, the graphene or reduced graphene layers help reduce corrosion [23]. Kumar et al. compared the catalytic activity of MoS_2/graphene oxide and MoS_2/reduced graphene oxide heterostructures and realized the higher activity of MoS_2/graphene oxide than MoS_2/reduced graphene oxide [119]. Reduction of functional groups of oxygen in graphene oxide results in reduced graphene oxide [120]. The path of electron movement in the MoS_2-graphene oxide composite may be easier than the MoS_2-reduced graphene oxide composite [4]. Table 4.7 shows a reported list of electrocatalysts in this category.

4.8 CONCLUSION

The importance of green hydrogen, HER mechanisms, and the role of nano-electrocatalysts in hydrogen production are discussed in detail in this chapter. The reaction mechanisms of HER, the role of activity parameters, and their measurements are crucial in controlling the production of green hydrogen. Various HER mechanisms such as Volmer, Heyrovsky, and Tafel reaction mechanisms, as well as RDS and its relationship with Tafel slope, are discussed. A detailed explanation is provided about the different HER activity parameters that can be derived from linear sweep voltammetry, CV, and EIS. The selection of the most suitable nano-electrocatalyst for hydrogen synthesis is a critical factor in the kinetics of the HER mechanism. The efficiency of a catalyst is crucial in the evolution of HER, as it is determined by the balanced rate of hydrogen adsorption and desorption on the catalyst's surface. Therefore, the design of a suitable nano-electrocatalyst that can enhance HER with a high rate of both adsorption and desorption processes is a challenging goal. This chapter provides a brief overview of the most popular nano-electrocatalysts and their HER activity. It also explores the role of nano-electrocatalysts in green hydrogen production. In conclusion, this chapter discusses the role of nano-electrocatalysts in green hydrogen production and the mechanisms of HER.

CONFLICT OF INTEREST

The authors declare no conflict of interest.

TABLE 4.7 Electrocatalytic Activity Parameters of Transition Metal Dichalcogenides/Carbon Derivative Composites

Electrocatalyst	**Overpotential at Current Density (mV)@mA/cm^2**	**Tafel Slope (mV/Dec)**	**TOF at Overpotential (s^{-1}) @mV**	**Durability (Cycles) or Hours**	**Capacitance (mF/cm^2)**	**Charge Transfer Resistance (Ω)**	**References**
CoS_2@MoS_2/rGO	98	37.4	/	2,000 cycles	23.04	/	[118]
MoS_2/GO	200 @10	40	/	15,000 S	0.253	/	[119]
Ni-MoS_2/GO	33 @10	110	/	2,000 cycles	77	0.94	[116]
MoS_2/GO-CNT	150 @74.25	38	/	1,000 cycles	/	/	[102]
MoS_2 nanoflower decorated rGO	10 @10	95	/	300 cycles	/	/	[96]
WS_2/rGO	300 @7	58	/	/	/	/	[81]
V doped MoS_2/rGO	262 @10	71	7.47 @300	1,000 cycles	/	3.384	[44]

KEYWORDS

- **carbon derivative heterostructures**
- **electrocatalysts**
- **green hydrogen**
- **hydrogen evolution reaction**
- **hydrogen fuel cell**
- **transition metal chalcogenides**
- **transition metal phosphides**

REFERENCES

1. IEA, (2022). *Global Hydrogen Review 2022*. IEA, Paris https://www.iea.org/reports/global-hydrogen-review-2022,License:CCBY4.0 (accessed on 16 January 2023).
2. Zou, X., & Zhang, Y., (2015). Noble metal-free hydrogen evolution catalysts for water splitting. *Chemical Society Reviews, 44*(15), 5148–5180.
3. Dincer, I., (2012). Green methods for hydrogen production. *International Journal of Hydrogen Energy, 37*(2), 1954–1971.
4. Benck, J. D., Hellstern, T. R., et al., (2014). Catalyzing the hydrogen evolution reaction (HER) with molybdenum sulfide nanomaterials. *ACS Catalysis, 4*(11), 3957–3971.
5. Liu, W., Xia, T., et al., (2021). Self-supported Ni_3N nanoarray as an efficient nonnoble-metal catalyst for alkaline hydrogen evolution reaction. *International Journal of Hydrogen Energy, 46*(53), 27037–27043.
6. McWhorter, S., Read, C., et al., (2011). Materials-based hydrogen storage: Attributes for near-term, early market PEM fuel cells. *Current Opinion in Solid State and Materials Science, 15*(2), 29–38.
7. Wilberforce, T., Alaswad, A., et al., (2016). Advances in stationary and portable fuel cell applications. *International Journal of Hydrogen Energy, 41*(37), 16509–16522.
8. Harrag, A., & Messalti, S., (2018). How fuzzy logic can improve PEM fuel cell MPPT performances? *International Journal of Hydrogen Energy, 43*(1), 537–550.
9. Balkan Green Energy News, (2022). *China Building World's Biggest Green Hydrogen Factory*. [Internet]. Available from: https://balkangreenenergynews.com (accessed on 16 January 2023).
10. Lubitz, W., & Tumas, W., (2007). Hydrogen: An overview. *Chemical Reviews, 107*(10), 3900–3903.
11. Abad, A. V., & Dodds, P. E., (2020). Green hydrogen characterization initiatives: Definitions, standards, guarantees of origin, and challenges. *Energy Policy, 138*, 111300.
12. Dincer, I., & Acar, C., (2018). Smart energy solutions with hydrogen options. *International Journal of Hydrogen Energy, 43*(18), 8579–8599.

13. Acar, C., & Dincer, I., (2018). 3.1 hydrogen production. *Comprehensive Energy Systems, 3*, 1–40.
14. Dawood, F., Anda, M., et al., (2020). Hydrogen production for energy: An overview. *International Journal of Hydrogen Energy, 45*(7), 3847–3869.
15. Zhang, Q., Uchaker, E., et al., (2013). Nanomaterials for energy conversion and storage. *Chemical Society Reviews, 42*(7), 3127–3171.
16. Wang, X., Jin, S., et al., (2015). Relation between the photocatalytic and photoelectrocatalytic performance for the particulate semiconductor-based photoconversion systems with surface phase junction structure. *The Journal of Physical Chemistry C, 119*(39), 22460–22464.
17. Liu, Y., Zhang, H., et al., (2018). 0D (MoS_2)/2D (g-C_3N_4) heterojunctions in Z-scheme for enhanced photocatalytic and electrochemical hydrogen evolution. *Applied Catalysis B: Environmental, 228*, 64–74.
18. Ross, D. K., (2006). Hydrogen storage: The major technological barrier to the development of hydrogen fuel cell cars. *Vacuum, 80*(10), 1084–1089.
19. Howarth, R. W., & Jacobson, M. Z., (2021). How green is blue hydrogen? *Energy Science & Engineering, 9*(10), 1676–1687.
20. Alghassab, M. A. (2018). *Non-Linear Control of Chaotic Forced Oscillators: Renewable Energy Application*. Doctoral dissertation, Oakland University.
21. Ueckerdt, F., Bauer, C., et al., (2021). Potential and risks of hydrogen-based e-fuels in climate change mitigation. *Nature Climate Change, 11*(5), 384–393.
22. Wu, A., Xie, Y., et al., (2018). Integrating the active OER and HER components as the heterostructures for the efficient overall water splitting. *Nano Energy, 44*, 353–363.
23. Wang, J., Xu, F., et al., (2017). Non-noble metal-based carbon composites in hydrogen evolution reaction: Fundamentals to applications. *Advanced Materials, 29*(14), 1605838.
24. Lasia, A., (2010). Hydrogen evolution reaction. *Handbook of Fuel Cells, 815*.
25. Lv, C., Xu, S., et al., (2019). Promoting electrocatalytic activity of cobalt cyclotetraphosphate in full water splitting by titanium-oxide-accelerated surface reconstruction. *Journal of Materials Chemistry A, 7*(20), 12457–12467.
26. Al-Ghussain, L., et al., (2022). Exploring the feasibility of green hydrogen production using excess energy from a country-scale 100% solar-wind renewable energy system. *International Journal of Hydrogen Energy*.
27. Nasser, M., Megahed, T. F., et al., (2022). A review of water electrolysis–based systems for hydrogen production using hybrid/solar/wind energy systems. *Environmental Science and Pollution Research*, 1–25.
28. Hwang, J. J., et al., (2009). Dynamic modeling of a photovoltaic hydrogen fuel cell hybrid system. *International Journal of Hydrogen Energy, 34*(23), 9531–9542.
29. Sunitha, A. P., (2021). *Article of Hydrogen: Energy Source of Future*. Yojana.
30. Cheng, X., Shi, Z., et al., (2007). A review of PEM hydrogen fuel cell contamination: Impacts, mechanisms, and mitigation. *Journal of Power Sources, 165*(2), 739–756.
31. Vishnyakov, V. M., (2006). Proton exchange membrane fuel cells. *Vacuum, 80*(10), 1053–1065.
32. Mori, D., & Hirose, K., (2009). Recent challenges of hydrogen storage technologies for fuel cell vehicles. *International Journal of Hydrogen Energy, 34*(10), 4569–4574.
33. Venturi, M., Mohrdieck, C., & Friedrich, J., (2013). Mercedes-Benz B-class fuel cell: The world largest hydrogen vehicle fuel cell fleet experience. In: 2013 *World Electric Vehicle Symposium and Exhibition (EVS27)* (pp. 1–11). IEEE.

34. Frenette, G., & Forthoffer, D., (2009). Economic & commercial viability of hydrogen fuel cell vehicles from an automotive manufacturer perspective. *International Journal of Hydrogen Energy, 34*(9), 3578–3588.
35. Li, M., Bai, Y., et al., (2019). Review on the research of hydrogen storage system fast refueling in fuel cell vehicle. *International Journal of Hydrogen Energy, 44*(21), 10677–10693.
36. Brown, L. F., (2001). A comparative study of fuels for on-board hydrogen production for fuel-cell-powered automobiles. *International Journal of Hydrogen Energy, 26*(4), 381–397.
37. Council, H., (2020). *The Hydrogen Council–An Introduction*. Belgium: Hydrogen Council.
38. Behling, N., Williams, M. C., et al., (2015). Fuel cells and the hydrogen revolution: Analysis of a strategic plan in Japan. *Economic Analysis and Policy, 48*, 204–221.
39. Dincer, I., Colpan, C. O., et al., (2013). *Causes, Impacts and Solutions to Global Warming*. Springer Science & Business Media.
40. Xiao, W., Liu, P., et al., (2017). Dual-functional N dopants in edges and basal plane of MoS_2 nanosheets toward efficient and durable hydrogen evolution. *Advanced Energy Materials, 7*(7), 1602086.
41. Xu, W., Li, S., et al., (2018). Large dendritic monolayer MoS_2 grown by atmospheric pressure chemical vapor deposition for electrocatalysis. *ACS Applied Materials & Interfaces, 10*(5), 4630–4639.
42. Joy, J., Mathew, J., & George, S. C., (2018). Nanomaterials for photoelectrochemical water splitting–review. *International Journal of Hydrogen Energy, 43*(10), 4804–4817.
43. Styring, S., (2012). Artificial photosynthesis for solar fuels. *Faraday Discussions, 155*, 357–376.
44. Singh, A. K., et al., (2019). Vanadium doped few-layer ultrathin MoS_2 nanosheets on reduced graphene oxide for high-performance hydrogen evolution reaction. *RSC Advances, 9*(39), 22232–22239.
45. Bai, S., et al., (2015). Chemically exfoliated metallic MoS_2 nanosheets: A promising supporting co-catalyst for enhancing the photocatalytic performance of TiO_2 nanocrystals. *Nano Research, 8*(1), 175–183.
46. Guo, X., et al., (2013). High photoelectrocatalytic performance of a MoS_2–SiC hybrid structure for hydrogen evolution reaction. *Journal of Materials Chemistry A, 1*(15), 4657–4661.
47. Paulraj, G., et al., (2020). Stable and highly efficient MoS_2/Si NWs hybrid heterostructure for photoelectrocatalytic hydrogen evolution reaction. *International Journal of Hydrogen Energy, 45*(3), 1793–1801.
48. Zhang, S., Li, X., et al., (2020). Enhancement of the photoelectrocatalyticH_2 evolution on a rutile-TiO_2 (001) surface decorated with dendritic MoS_2 monolayer nanoflakes. *ACS Applied Energy Materials, 3*(6), 5756–5764.
49. Paul, K. K., et al., (2018). Strongly enhanced visible light photoelectrocatalytic hydrogen evolution reaction in an n-doped MoS_2/TiO_2 (B) heterojunctions by selective decoration of platinum nanoparticles at the MoS_2 edge sites. *Journal of Materials Chemistry A, 6*(45), 22681–22696.
50. Koutavarapu, R., et al., (2020). Carbon cloth/transition metals-based hybrids with controllable architectures for electrocatalytic hydrogen evolution: A review. *International Journal of Hydrogen Energy, 45*(13), 7716–7740.

51. Majhi, K. C., & Yadav, M., (2020). Transition metal chalcogenides-based nanocomposites as efficient electrocatalyst for hydrogen evolution reaction over the entire pH range. *International Journal of Hydrogen Energy, 45*(46), 24219–24231.
52. Gong, M., Wang, D. Y., et al., (2016). A mini review on nickel-based electrocatalysts for alkaline hydrogen evolution reaction. *Nano Research, 9*(1), 28–46.
53. Tian, X., Zhao, P., & Sheng, W., (2019). Hydrogen evolution and oxidation: Mechanistic studies and material advances. *Advanced Materials, 31*(31), 1808066.
54. Ge, Z., et al., (2020). A review of the electrocatalysts on hydrogen evolution reaction with an emphasis on Fe, Co and Ni-based phosphides. *Journal of Materials Science, 55*(29), 14081–14104.
55. Hua, W., et al., (2020). A review and perspective on molybdenum-based electrocatalysts for hydrogen evolution reaction. *Rare Metals, 39*(4), 335–351.
56. Fu, Q., et al., (2021). 2D transition metal dichalcogenides: Design, modulation, and challenges in electrocatalysis. *Advanced Materials, 33*(6), 1907818.
57. Cheng, Y., et al., (2017). Os/Si nanocomposites as excellent hydrogen evolution electrocatalysts with thermodynamically more favorable hydrogen adsorption free energy than platinum. *Nano Energy, 39*, 284–290.
58. Xiao, P., Chen, W., & Wang, X., (2015). A review of phosphide-based materials for electrocatalytic hydrogen evolution. *Advanced Energy Materials, 5*(24), 1500985.
59. Nørskov, J. K., et al., (2005). Trends in the exchange current for hydrogen evolution. *Journal of The Electrochemical Society, 152*(3), J23.
60. Yu, P., et al., (2019). Earth abundant materials beyond transition metal dichalcogenides: A focus on electro catalyzing hydrogen evolution reaction. *Nano Energy, 58*, 244–276.
61. Zhou, Q., et al., (2020). A feasible and environmentally friendly method to simultaneously synthesize MoS_2 quantum dots and pore-rich monolayer MoS_2 for hydrogen evolution reaction. *International Journal of Hydrogen Energy, 45*(1), 433–442.
62. Sharma, R., et al., (2019). On the synthesis of morphology-controlled transition metal dichalcogenides via chemical vapor deposition for electrochemical hydrogen generation. *Physica Status Solidi (RRL)–Rapid Research Letters, 13*(12), 1900257.
63. Xie, J., et al., (2013). Controllable disorder engineering in oxygen-incorporated MoS_2 ultrathin nanosheets for efficient hydrogen evolution. *Journal of the American Chemical Society, 135*(47), 17881–17888.
64. Debata, S., et al., (2019). Marigold shaped N-rGO-MoS_2-Ni $(OH)_2$ nanocomposite as a bifunctional electrocatalyst for the promotion of overall water splitting in alkaline medium. *Electrochimica Acta, 303*, 257–267.
65. Benck, J. D., et al., (2014). Catalyzing the hydrogen evolution reaction (HER) with molybdenum sulfide nanomaterials. *ACS Catalysis, 4*(11), 3957–3971.
66. Saha, A., et al., (2020). Exfoliated colloidal MoS_2 nanosheet with predominantly 1T phase for electrocatalytic hydrogen production. *International Journal of Hydrogen Energy, 45*(37), 18645–18656.
67. Ali, A., & Shen, P. K., (2020). Nonprecious metal's graphene-supported electrocatalysts for hydrogen evolution reaction: Fundamentals to applications. *Carbon Energy, 2*(1), 99–121.
68. Gao, Q., et al., (2019). Structural design and electronic modulation of transition-metal-carbide electrocatalysts toward efficient hydrogen evolution. *Advanced Materials, 31*(2), 1802880.

69. Sunitha, A. P., et al., (2021). Hydrothermally grown MoS_2 quantum dots for electrocatalytic applications. In: *AIP Conference Proceedings* (Vol. 2379, No. 1, p. 030003). AIP Publishing LLC.
70. Zhang, J., et al., (2016). Engineering water dissociation sites in MoS_2 nanosheets for accelerated electrocatalytic hydrogen production. *Energy & Environmental Science, 9*(9), 2789–2793.
71. Hu, J., Zhang, C., et al., (2020). Kinetic-oriented construction of MoS_2 synergistic interface to boost pH-universal hydrogen evolution. *Advanced Functional Materials, 30*(6), 1908520.
72. Ekspong, J., et al., (2018). Stable sulfur-intercalated 1T′ MoS_2 on graphitic nanoribbons as hydrogen evolution electrocatalyst. *Advanced Functional Materials, 28*(46), 1802744.
73. Sun, K., et al., (2018). Targeted bottom-up synthesis of 1T-phase MoS_2 arrays with high electrocatalytic hydrogen evolution activity by simultaneous structure and morphology engineering. *Nano Research, 11*(8), 4368–4379.
74. Wang, H., Thiele, A., & Pilon, L., (2013). Simulations of cyclic voltammetry for electric double layers in asymmetric electrolytes: A generalized modified Poisson–Nernst–Planck model. *The Journal of Physical Chemistry C, 117*(36), 18286–18297.
75. Wang, H., & Pilon, L., (2012). Physical interpretation of cyclic voltammetry for measuring electric double layer capacitances. *Electrochimica Acta, 64*, 130–139.
76. Wang, H., & Pilon, L., (2011). Accurate simulations of electric double layer capacitance of ultramicroelectrodes. *The Journal of Physical Chemistry C, 115*(33), 16711–16719.
77. Bard, A. J., & Faulkner, L. R., (2001). Fundamentals and applications: Electrochemical methods. *Electrochem. Methods, 2*(482), 580–632.
78. Krishnamoorthy, K., et al., (2014). Supercapacitive properties of hydrothermally synthesized sphere like MoS_2 nanostructures. *Materials Research Bulletin, 50*, 499–502.
79. Soon, J. M., & Loh, K. P., (2007). Electrochemical double-layer capacitance of MoS_2 nanowall films. *Electrochemical and Solid-State Letters, 10*(11), A250.
80. Girard, H. L., et al., (2015). Physical interpretation of cyclic voltammetry for hybrid pseudocapacitors. *The Journal of Physical Chemistry C, 119*(21), 11349–11361.
81. Yang, J., Voiry, D., et al., (2013). Two-dimensional hybrid nanosheets of tungsten disulfide and reduced graphene oxide as catalysts for enhanced hydrogen evolution. *Angewandte Chemie, 125*(51), 13996–13999.
82. Nayana, K., et al., (2021). Probing nanostructure MoS_2 as catalyst in light activated and electro activated hydrogen evolution reaction. *Materials Today: Proceedings, 45*, 4671–4676.
83. Yao, Y., et al., (2019). MoS_2 coexisting in 1T and 2H phases synthesized by common hydrothermal method for hydrogen evolution reaction. *Nanomaterials, 9*(6), 844.
84. Saseendran, S. B., et al., (2020). Hydrothermal synthesis of MoS_2 for supercapacitive application. In: *AIP Conference Proceedings* (Vol. 2244, No. 1, p. 070034). AIP Publishing LLC.
85. Yang, J., & Shin, H. S., (2014). Recent advances in layered transition metal dichalcogenides for hydrogen evolution reaction. *Journal of Materials Chemistry A, 2*(17), 5979–5985.
86. Wu, L., et al., (2017). Mn doped MoS_2/reduced graphene oxide hybrid for enhanced hydrogen evolution. *Applied Surface Science, 425*, 470–477.
87. Chen, Z., et al., (2011). Core–shell MoO_3–MoS_2 nanowires for hydrogen evolution: A functional design for electrocatalytic materials. *Nano Letters, 11*(10), 4168–4175.

88. Xie, J., et al., (2013). Defect-rich MoS_2 ultrathin nanosheets with additional active edge sites for enhanced electrocatalytic hydrogen evolution. *Advanced Materials, 25*(40), 5807–5813.
89. Peng, X., et al., (2020). Recent advance and prospective of electrocatalysts based on transition metal selenides for efficient water splitting. *Nano Energy, 78*, 105234.
90. Premnath, K., et al., (2019). Hydrothermally synthesized nickel molybdenum selenide composites as cost-effective and efficient trifunctional electrocatalysts for water splitting reactions. *International Journal of Hydrogen Energy, 44*(41), 22796–22805.
91. Zhu, M., et al., (2021). Iron molybdenum selenide supported on reduced graphene oxide as an efficient hydrogen electrocatalyst in acidic and alkaline media. *Journal of Colloid and Interface Science, 602*, 384–393.
92. Wang, X. L., et al., (2019). Molecular modulation of a molybdenum–selenium cluster by sulfur substitution to enhance the hydrogen evolution reaction. *Inorganic Chemistry, 58*(18), 12415–12421.
93. Zhou, Y. N., et al., (2020). Optimized Mo–doped cobalt selenides coupled carbon nanospheres for efficient hydrogen evolution. *Applied Surface Science, 531*, 147404.
94. Xia, C., et al., (2017). Active edge sites engineering in nickel cobalt selenide solid solutions for highly efficient hydrogen evolution. *Advanced Energy Materials, 7*(9), 1602089.
95. Boakye, F. O., et al., (2022). One-step synthesis of heterostructured cobalt-iron selenide as bifunctional catalyst for overall water splitting. *Materials Chemistry and Physics, 275*, 125201.
96. Ma, C. B., Qi, X., et al., (2014). MoS_2 nanoflower-decorated reduced graphene oxide paper for high-performance hydrogen evolution reaction. *Nanoscale, 6*(11), 5624–5629.
97. Kadam, S. R., et al., (2021). Facile synthetic approach to produce optimized molybdenum carbide catalyst for alkaline HER. *Applied Surface Science, 559*, 149932.
98. Liu, G., et al., (2019). A highly efficient alkaline HER Co–Mo bimetallic carbide catalyst with an optimized Mo d-orbital electronic state. *Journal of Materials Chemistry A, 7*(20), 12434–12439.
99. Wan, C., et al., (2014). Multiple phases of molybdenum carbide as electrocatalysts for the hydrogen evolution reaction. *Angewandte Chemie, 126*(25), 6525–6528.
100. Cui, Y., et al., (2020). Tungsten oxide/carbide surface heterojunction catalyst with high hydrogen evolution activity. *ACS Energy Letters, 5*(11), 3560–3568.
101. Yang, C., et al., (2022). Balance effect: A universal strategy for transition metal carbides to enhance hydrogen evolution. *Advanced Functional Materials, 32*(5), 2108167.
102. Khan, M., Yousaf, A. B., et al., (2016). Molybdenum sulfide/graphene-carbon nanotube nanocomposite material for electrocatalytic applications in hydrogen evolution reactions. *Nano Research, 9*(3), 837–848.
103. Xu, K., et al., (2017). Regulating water-reduction kinetics in cobalt phosphide for enhancing HER catalytic activity in alkaline solution. *Advanced Materials, 29*(28), 1606980.
104. Wang, X., et al., (2015). One-step synthesis of self-supported nickel phosphide nanosheet array cathodes for efficient electrocatalytic hydrogen generation. *Angewandte Chemie, 127*(28), 8306–8310.
105. Popczun, E. J., et al., (2013). Nanostructured nickel phosphide as an electrocatalyst for the hydrogen evolution reaction. *Journal of the American Chemical Society, 135*(25), 9267–9270.

106. Wang, M., et al., (2020). Surface engineering by doping manganese into cobalt phosphide towards highly efficient bifunctional HER and OER electrocatalysis. *Applied Surface Science, 515*, 146059.
107. Xiao, P., et al., (2014). Molybdenum phosphide as an efficient electrocatalyst for the hydrogen evolution reaction. *Energy & Environmental Science, 7*(8), 2624–2629.
108. Tian, L., et al., (2016). Electrochemical activity of iron phosphide nanoparticles in hydrogen evolution reaction. *ACS Catalysis, 6*(8), 5441–5448.
109. Chang, B., et al., (2018). Bimetallic NiMoN nanowires with a preferential reactive facet: An ultraefficient bifunctional electrocatalyst for overall water splitting. *Chem Sus Chem, 11*(18), 3198–3207.
110. Xiong, J., et al., (2017). Salt-templated synthesis of defect-rich MoN nanosheets for boosted hydrogen evolution reaction. *Journal of Materials Chemistry A, 5*(46), 24193–24198.
111. Yan, H., et al., (2015). Phosphorus-modified tungsten nitride/reduced graphene oxide as a high-performance, non-noble-metal electrocatalyst for the hydrogen evolution reaction. *Angewandte Chemie, 127*(21), 6423–6427.
112. Wang, Z., et al., (2019). Copper–nickel nitride nanosheets as efficient bifunctional catalysts for hydrazine-assisted electrolytic hydrogen production. *Advanced Energy Materials, 9*(21), 1900390.
113. Zhu, Y., Lin, Q., et al., (2020). Metal oxide-based materials as an emerging family of hydrogen evolution electrocatalysts. *Energy & Environmental Science, 13*(10), 3361–3392.
114. Li, Y. H., et al., (2015). Local atomic structure modulations activate metal oxide as electrocatalyst for hydrogen evolution in acidic water. *Nature Communications, 6*(1), 1–7.
115. Zhang, L., et al., (2018). N-modified NiO surface for superior alkaline hydrogen evolution. *Chem Sus Chem, 11*(6), 1020–1024.
116. Yin, X., Sun, G., et al., (2017). A novel structure of Ni-(MoS_2/GO) composite coatings deposited on Ni foam under supergravity field as efficient hydrogen evolution reaction catalysts in alkaline solution. *Electrochimica Acta, 249*, 52–63.
117. Song, C., Zhao, Z., et al., (2019). *In situ* growth of Ag nanodots decorated Cu_2O porous nanobelts networks on copper foam for efficient HER electrocatalysis. *Small, 15*(29), 1804268.
118. Guo, Y., Gan, L., et al., (2017). A cake-style CoS_2@MoS_2/RGO hybrid catalyst for efficient hydrogen evolution. *Advanced Functional Materials, 27*(5), 1602699.
119. Kumar, S., et al., (2017). Electrochemical and SECM investigation of MoS_2/GO and MoS_2/rGO nanocomposite materials for HER electrocatalysis. *ACS Omega, 2*(11), 7532–7545.
120. Stobinski, L., Lesiak, B., et al., (2014). Graphene oxide and reduced graphene oxide studied by the XRD, TEM and electron spectroscopy methods. *Journal of Electron Spectroscopy and Related Phenomena, 195*, 145–154.

PART II
Smart Technologies

CHAPTER 5

Reengineering Processes of Youth Exposure on Agriculture and Environmental Sustainability: Revealing Big Data Technology and Security Adoption in the Digital Ecosystem

HERU SUSANTO,[1–3] ARIF JUNUIN,[1] and ALIFYA KAYA SHAFA[2,3]

[1]*Center for Innovative Engineering and School of Business, Universiti Teknologi Brunei, Mukim Gadong A, Brunei*

[2]*Center for Research Collaboration of Graph Theory and Combinatory, UTB Brunei Darussalam-BRIN Indonesia-THU Taiwan-ITB Indonesia-UI Indonesia, and National Research and Innovation Agency, Jakarta, Indonesia*

[3]*Department of Computer Science, Tunghai University, Taichung City, Taiwan*

ABSTRACT

This research study aims to analyze and study the exposure of youth in the agricultural field and understand the extent to which they are exposed to it. The researcher intends to explore the youth's perception of the agricultural field, their interest in working in the agriculture sector, and identify technological innovations that can sustain and support the agricultural field. The research will begin by analyzing the objectives and needs of the study. The methodology will involve collecting primary data through interviews and questionnaire surveys, as well as utilizing secondary data from research

Sustainability in Energy and Environment: Engineered Materials and Smart Computational Techniques. Ann Rose Abraham, Heru Susanto, A. K. Haghi, & Kian Hariri Asli (Eds.)

papers, journals, and articles. Statistical analysis, including One-sample t-test analysis and Descriptive analysis, will be used to analyze the data from the questionnaire survey and test the hypotheses for significance. The questionnaire will be used to gather public perceptions of the agricultural field, aligning with one of the research objectives. During a seminar titled "The Involvement of the Society, Specifically Youth in the Agricultural Sector," it was highlighted that there is a lack of interest in agriculture among the youth in Brunei, as well as an urban perception of cultivation. This research aims to discuss the increasing unemployment rate, the causes of decline in youth participation in agricultural work, actions that can be taken by the youth to support agricultural work, techniques to attract youth to the agriculture field, problems that Brunei faces in terms of agriculture, and technological innovations that can sustain and support agriculture. The main focus is to determine the extent of youth exposure to the agricultural field. All individuals, government, and non-government sectors play an important role in raising awareness of the agriculture sector.

5.1 INTRODUCTION

5.1.1 BACKGROUND OF THE STUDY

Agriculture has become an important economic activity in Brunei, but its significance has declined since the discovery of oil in 1929. In 1947, over 50% of the population was employed in agriculture, but by 1971, this number had decreased to 10%. In the early 1960s, there was concern about the economic consequences of over-reliance on oil, and the Second National Development Plan of 1962–1968 focused on agriculture [1]. According to the Department of Government Printing, the office of the Prime Minister, Brunei Darussalam is a small country with abundant opportunities. It has slightly less than half a million inhabitants, yet it ranks among the countries with the highest macroeconomic stability rates in the world. The current population of Negara Brunei Darussalam is 437,436, based on Worldometer's elaboration of the latest United Nations data. The population of Brunei Darussalam is equivalent to 0.01% of the total world population. Brunei ranks 175th in the list of countries and dependencies by population. Brunei is one of Southeast Asia's richest economies, and there is no denying the fact that oil and gasoline sales have contributed tremendously to the way of life in Brunei Darussalam. The location of Brunei Darussalam is northwest of the island of Borneo, between latitudes 4° and 5°5' North and longitudes 11.4°4' and 11°23' East.

The land area is about 5,765 square kilometers. The climate of Brunei has a temperature range between 23.1°C and 32.0°C, with relative humidity and total rainfall of 1386.1 millimeters. Figure 5.1 shows Brunei's map.

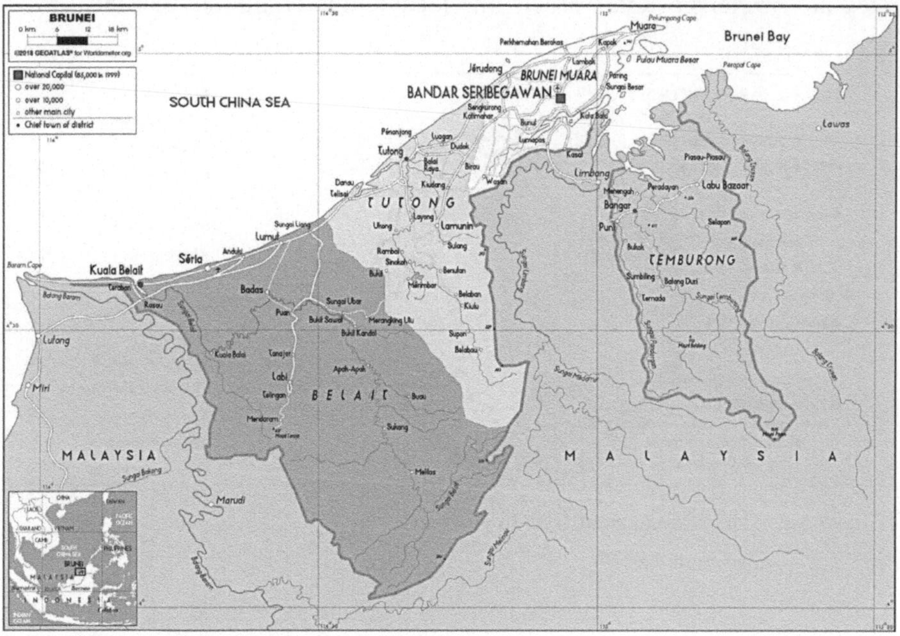

FIGURE 5.1 The map of Brunei Darussalam.

Source: Reprinted from Ref. [8].

Throughout 2017, the Borneo Bulletin reported that oil production decreased the total inventory and reserves to 134,000 barrels per day compared to 2016, 139,000 barrels per day [2]. Brunei Darussalam is heavily reliant on oil and gas sales to finance a wide range of national and personal activities. According to the Department of Economics, Planning, and Development, oil and gas contributed 63.3% of GDP in 2014, as shown in Figure 5.2. Brunei Darussalam has been making consistent efforts to diversify its economy in line with the emerging trends in ASEAN. Policymakers consider small businesses to play a primary role [2].

Brunei has been taking actions to stimulate economic growth with the aid of The National Vision of Wawasan Brunei 2035. It has three predominant visions for Brunei Darussalam: to make Brunei Darussalam a nation widely recognized for the accomplishment of its educated and highly skilled people, as measured by the very best international standards; to achieve a quality of life that ranks among the top 10 countries in the world; and to build a dynamic and sustainable economic system with a per capita income among the world's top 10. One of the main objectives of Wawasan Brunei is to integrate Brunei's primary assets into local and international markets.

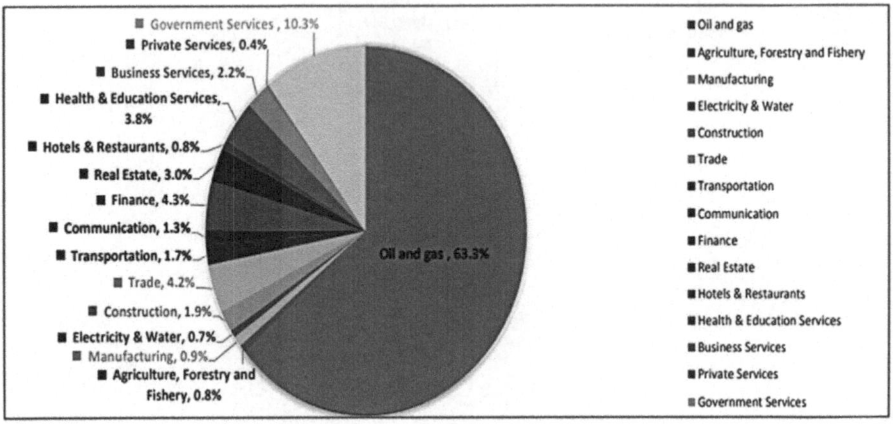

FIGURE 5.2 The percentage of GDP in 2014.

Source: Reprinted from Ref. [9].

The GDP of the Department of Economic Planning and Development is calculated based on stable prices, which are adjusted for inflation. These prices are determined by the economic system over a specific year. On the other hand, current prices reflect the market value of goods and services. Without adjusting for inflation, Brunei's highest GDP in the past decades was recorded in 2012 at US $19.048 billion. However, it declined for four consecutive years to US $11.401 billion in 2016, mainly due to a decrease in oil and gas prices. As oil and gas prices began to recover, Brunei's economy grew by 1.3% to US $12.28 billion in 2017, and further increased to US $13.567 billion in 2018, according to data from the World Bank as shown in Figure 5.3.

As can be seen in Figure 5.4, the agriculture GDP in Brunei Darussalam increased to 36.20 BND million in the second quarter of 2019 from 35.80 BND million in the first half. The GDP from agriculture in Brunei accumulated to BND 37.73 million between 2013 and 2018, reaching an all-time high of BND 42 million in the third quarter of 2015 and a low of BND 34.30 million in the fourth quarter of 2013.

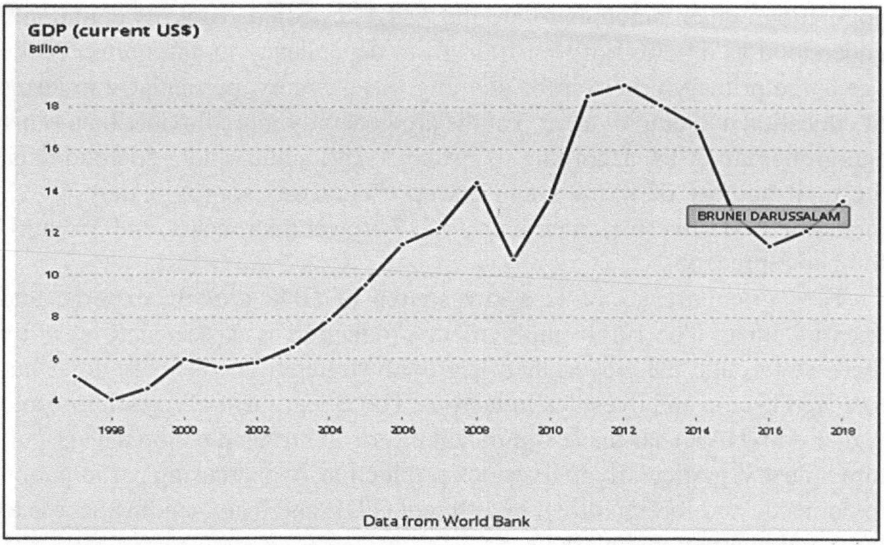

FIGURE 5.3 World Bank's GDP.

Source: Reprinted from Ref. [10].

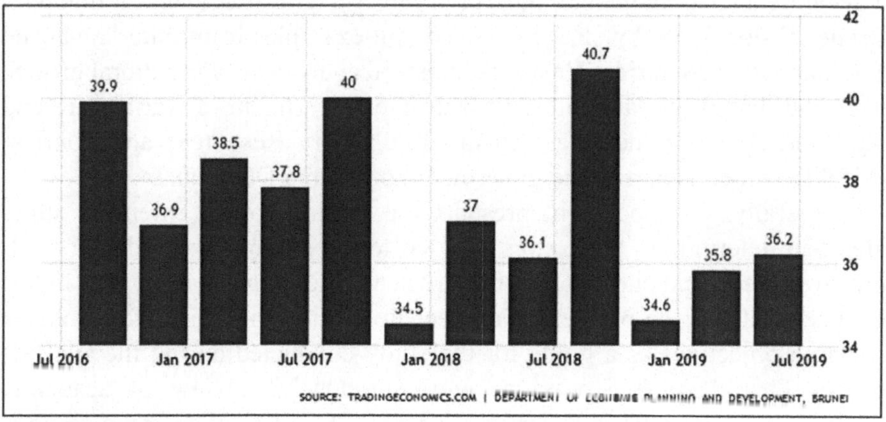

FIGURE 5.4 Brunei's GDP for agriculture.

Source: Reprinted from Ref. [11].

According to the Population Pyramid, Brunei Darussalam is expected to have a population of 445,249 in 2019. As stated by the Department of Economic Planning and Development, the total population of youths in Brunei, defined as individuals aged 15 to 40, is approximately 41% of the total population, which amounts to around 184,477 people. Youth is commonly understood as a period of transition from dependency to autonomy. While age is the primary criterion for defining this category, particularly in terms of education and employment, youths are generally more flexible than other age groups. In 2018, according to Brunei's Agriculture and Agri-food data, the total number of farmers and entrepreneurs was approximately 5,433, including 8,311 livestock marketers, 4,217 crop entrepreneurs, and 385 agri-food entrepreneurs.

The agriculture sector is a key source of GDP growth, exports, and external trade. The rural employment production is expected to steadily increase through the Government's growth strategy. The primary contributors are the crop and livestock industries. The Department of Agriculture and Agri-Food (DAA) has made significant efforts to further develop the agriculture industry, particularly in livestock production, by increasing participation in domestic and foreign direct investment (FDI) and by expanding access to local and international markets.

The mission is to evaluate the agricultural practices in Brunei Darussalam and understand how young adults are involved in the agricultural sector. The objective is to promote sustainable economic diversification by enhancing and strengthening agriculture in Brunei Darussalam, in line with Brunei Vision 2035, by studying successful examples from other advanced and innovative countries. The researcher will analyze the agricultural growth in Brunei based on data collected from the Department of Agriculture and Agrifood (DAA) under the Ministry of Primary Resources and Tourism (MPRT), which is part of the government of Brunei Darussalam.

Generally, this document presents the findings of an extensive study aimed at determining the factors that led to the attempt to validate a well-prepared piece, and outlines its structure in a professional manner. The report begins by stating its problem statement in relation to Brunei, followed by research objectives as a guide for the study conducted during the analysis phase. This document provides a comprehensive review of academic literature to comprehend the research topic. It will provide some insights into the research topic, summarize it within the conceptual framework, and include a significant chapter on this study. The Methodology phase will describe in depth, starting from the literature review, the translation of the analysis method, and its contribution to the creation of study techniques for data collection. The structure of the study's layout and the methods used are crucial for obtaining the primary facts from which this study can draw conclusions about the effects of the data collection. Additionally, data analysis will also be conducted in this study to determine the results, based on an interpretation of the findings. This will be accomplished through an evaluation and limitation phase, where this study elaborates on the results of the analysis and highlights the implications of this report throughout the research process. Finally, this report concludes with a section summarizing the examiner's findings and providing recommendations for future research.

5.1.2 PROBLEM STATEMENTS

Brunei hopes to stimulate the progress of agricultural output to promote economic growth and expansion. Therefore, the involvement of children in the agricultural sector needs to be promoted. According to statistics from the Department of Statistics and the Department of Economic Planning and Development, Brunei's unemployment rate increased to 9.3% in 2017 from 6.9% in 2014. The high unemployment rate among young Bruneians remains a growing concern, attributed to the lack of educational qualifications, limited job opportunities, and selective and unmotivated applicants. It is crucial for the youth in Brunei to cultivate the right mindset and approach when seeking job opportunities, and to become more self-reliant and proactive in their job search efforts to address this issue in the country. An investigation is being conducted to assess the extent of adolescent exposure to the agricultural sector in Brunei Darussalam. This study aims to examine the unemployment rate in Brunei, the level of youth involvement in agriculture, and the overall performance of Brunei's agricultural sector, including productivity, as well as government and non-governmental support. Therefore, the researcher's objective in this study is to enhance the overall performance of the agricultural sector in Brunei Darussalam, both in terms of government and non-governmental efforts. To address the research problem, the researcher will examine the contribution of Brunei's agriculture to the country's gross domestic product, the sustainability of agriculture in Brunei, the most viable agricultural practices in the country, the impact of the agricultural sector on Brunei's economy, and the technologies that can be utilized in Brunei's agriculture, as well as the challenges related to resource scarcity.

5.1.3 RESEARCH QUESTIONS

Based on the above discussion on the hassle statement, the core research question of the study is to what extent the youth are involved in the agricultural sector in Brunei Darussalam. The specific research questions were asked to guide the discussion:

- What is the youth perception on agriculture field?
- How interested are young people in working in the agriculture sector?
- What kinds of technological innovations can sustain and support the agricultural field?
- What actions need to be taken to attract attention from youth?

5.1.4 RESEARCH OBJECTIVES

The main objective of this study is to examine the exposure of youth to the agricultural field in Brunei Darussalam. Additionally, it aims to connect the literature reviews with the research questions and test the hypothesis that will be elaborated on in the next section. The research aims to determine whether Brunei Darussalam offers ample opportunities for youth to engage in farming. Furthermore, it seeks to explore the reasons for the absence of youth in the agricultural sector and assess whether increased youth participation is necessary. In conducting this study, the researcher should also consider the technological advancements that can support and benefit the agricultural sector.

5.2 LITERATURE REVIEW

5.2.1 OVERVIEW OF BRUNEI'S AGRICULTURE

The MPRT was established on October 22nd, 2015. This demonstrates His Majesty's commitment to enhancing the growth of the primary resource sectors, particularly Agriculture and Agri-food, Fisheries, and Forestry in Brunei Darussalam. The development of these sectors will contribute significantly to sustainable economic growth and the realization of Brunei Vision 2035. The challenge for the Agriculture and Agri-Food department is to strengthen the growth of agriculture and agri-food-based industries through the use of advanced technology and an export-oriented approach. This includes promoting both local and foreign direct investment and focusing on high-yield production and processing. Their vision is to develop agriculture and agri-food industries based on increased productivity and advanced technology for export purposes.

Brunei's agriculture has seen a 13% increase. His Majesty has pledged to provide more farmland and support. The gross agriculture output in Brunei has risen from $379.65 million to $436.30 million in 2018, thanks to the growth in livestock and crops. During the Hari Raya 2018 event organized by the Sultan Haji Hassanal Bolkiah Foundation (YSHHB), His Majesty mentioned the importance of opening more farmland and providing support to further develop the industry. In 2018, statistics on the agricultural output of the Sultanate were released during the Hari Raya celebration, where His Majesty the Sultan and Yang Di-Pertuan of Brunei Darussalam highlighted the development of over 300 hectares of land by new farmers under a unique pilot project introduced in 2016.

The Minister of Primary Resources and Tourism has stated that Brunei's first home-grown hybrid rice, Titih, has the potential to become the country's highest-yielding variety, with a consistent yield of eight tons per hectare once mass plantation begins. Yang Berhormat Dato Seri Setia Hj Ali Apong mentioned that foreign experts may soon arrive to initiate commercial production of Titih seedlings, which will be distributed to local farmers in stages. According to Biz Brunei, Titih has demonstrated a yield of 800 paddy per hectare during trials, reaching a height of 1.07 meters and maturing in 115 days. Rice farms in Wasan and the upcoming site Kandol, which have more comprehensive irrigation systems, are expected to be the first to cultivate this variety on a large scale.

Another brand-new hybrid rice variety from Indonesia, specifically Sembada188, has been jointly developed by the DAA and PT. Bio-gene plantation. It is currently the highest-yielding and most widely planted variety in Brunei, capable of producing six tons per hectare. The well-known local variety, Laila, produces three tons per hectare. The national average yield in 2017, which includes small-scale farms planting traditional varieties once a year, was 175 tons per hectare. The mass plantation of Titih has been a long awaited development, with research on the strain dating back to 2011 with Sunland Pte Ltd. According to reports, breeding hybrid strains locally is a complex process that requires specialized expertise. It involves cross-pollination from specific parental lineages, which results in the modification of the reproductive systems of rice plants.

The nearby farmers face various challenges and demanding conditions, such as uncertain weather situations, geographical location, land use and land cover, expensive maintenance, lack of economic support, and competition from neighboring states in the market. The local agricultural region experiences high production costs, low productivity, and inferior product quality. Additionally, agricultural entrepreneurs in the vicinity must be prepared to face the challenges of trade liberalization in the ASEAN region and the increasingly open and competitive global market. Local farmers need to become self-reliant and transform the local agricultural sector into a market-driven agribusiness. One of the major challenges they face is pests in rice crops, which significantly affect production as the pests feed on the roots, stems, leaves, and grains of the paddy.

5.2.2 BRUNEI'S INITIATIVES AND STRATEGIES TOWARDS AGRICULTURE SECTOR

The farming pilot project under the DAA is one of nine initiatives developed by the MPRT to increase agricultural output to meet local demands as well as for export. According to Biz Brunei, approximately 46 new companies, including 16 owned by local youth, have started farming predominantly vegetables and fruits under the pilot project. These companies benefit from heavily subsidized land rental costs, which the DAA's Deputy Director, Hj Rusali Hj Sapar, stated are among the cheapest in ASEAN. Most of the companies participating in the pilot project are utilizing hundreds of hectares. In general, over 7,000 hectares are gazetted under KKP for agricultural use. MPRT Minister Yang Berhormat Dato Seri Setia Hj Ali Apong noted that the farmers' lease period under the pilot project could be extended if they demonstrate progress. Annual rental fees for KKP start at $5 per hectare for vegetables and livestock, and go up to $25 for fruit farming. The program is also open to foreigners residing nearby.

The Agriculture Commercial Agency Exposure Software Program for youngsters on a Contract Farming basis is aimed at attracting the interest of teenagers to venture into rural areas, which would help increase agricultural output and enhance the country's export market. The application is one of the initiatives of the Agriculture and Agri-food Department to support the department's 2016 to 2020 strategic plan. The matter was reported in 2018 by Dayang Hajah Aidah binti Haji Mohd Hanifah, Director of Agriculture and Agri-food, during a briefing on the program at the Rimba Horticulture Business Center in Gadong. According to Dayang Hajah Aidah, the program utilizes advanced technology, which can boost agricultural output and provide profitable returns. She added that the children will be exposed to real-world technology, which can enhance productivity through field visits that are included in the program. This program aims to attract interest and increase youth participation in the agricultural sector, based on a contract farming model, to boost agricultural output and enhance the country's export market. The briefing also discussed investment opportunities in crop management, as well as economic simulations for agricultural companies using advanced technology.

Increasing marketplace access and other obligations of MPRT aim to enhance farmers' access to markets through business matching, contract farming, and meeting international food safety standards to enable local produce to be exported. DAA Senior Agriculture Officer Hirman Hj Abu mentioned that major retailers such as Sim Kim Huat and Hua Ho have established contracts with local farmers for supply, while large farms like Hua Ho's agricultural arm and Pertanian Tropikal Utama are starting to collaborate with smaller farms to cultivate produce they believe they can market.

Figure 5.5 shows Brunei's Agricultural Gross Output. The Minister of Primary Resources and Tourism has deliberated on numerous new strategies to assist existing farmers, younger farmers, and low-income farmers. The Ministry emphasizes the importance of focusing on high-quality plantations with modern techniques such as the use of hydroponic systems, vertical farming, and greenhouses. The goal is to target both the domestic and export markets. For low-income farmers, affordable and simple techniques such as traditional and fertigation methods will be introduced, along with the use of profitable by-products. Additionally, the Ministry has engaged in discussions with various agricultural product sourcing and export companies, as well as several major supermarkets in the country, to establish contracts with local farmers who can meet the market's demand for both local and export markets. Referring to the development price range for the 2018 to 2019 period, there is an allocation of 1 million dollars. The cost of the plan amounts to 3 million dollars under the allocation of the 11th National Development Plan, primarily for the enhancement program of manufacturing vegetables using advanced technology. Additionally, the recurring budget allocates over 468,000 dollars to enhance the biosecurity of agriculture and ensure that agricultural products can reach the export market.

5.2.3 YOUTH UNEMPLOYMENT IN BRUNEI DARUSSALAM

Today, this phenomenon remains a worldwide problem to which Brunei Darussalam isn't always continuously immune. Despite being stated as one of the richest countries in the world with an exceedingly small population, as said in The World Factbook, Brunei faced an unemployment rate of 2.6% in 2011, creating close-by news headlines sometimes in ultra-present-day years.

Likewise, unemployment has precipitated several issues to rise socially, economically, and psychologically. One of the reasons for unemployment in Brunei is low qualifications. The eternal secretary of culture, youngsters, and sports, Dato Haji Mohd Hamid, stated that among the 18,000 locals unemployed, a massive variety of them are the ones with handiest number one or secondary diploma qualifications [3], which makes it hard for them to discover the ideal activity. The second purpose is due to assignment unavailability and financial instability. According to a study conducted by the Center for Strategic and Policy Studies (CSPS), at least 30,000 jobs need to be created to address this unemployment problem, especially within the private sector. However, the personal sector in Brunei is not massive enough to create extra jobs. In addition, they find the kingdom of the economy adverse to the growth and employment of large portions of labor. Having discovered low job opportunities and little economic improvement over the years, graduates feel that possibilities that are available someplace else, which is in other international locations, are well worth taking, and scholarship students analyzing abroad may feel disinclined to return home and serve the country.

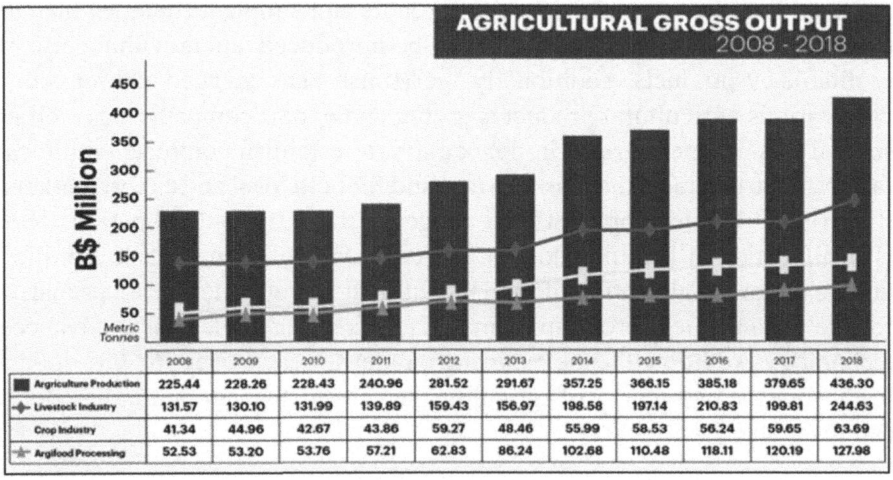

	2008	2009	2010	2011	2012	2013	2014	2015	2016	2017	2018
Argriculture Production	225.44	228.26	228.43	240.96	281.52	291.67	357.25	366.15	385.18	379.65	436.30
Livestock Industry	131.57	130.10	131.99	139.89	159.43	156.97	198.58	197.14	210.83	199.81	244.63
Crop Industry	41.34	44.96	42.67	43.86	59.27	48.46	55.99	58.53	56.24	59.65	63.69
Argifood Processing	52.53	53.20	53.76	57.21	62.83	86.24	102.68	110.48	118.11	120.19	127.98

FIGURE 5.5 Agricultural gross output.

Source: Reprinted from Ref. [12].

The final factor to which the persistence of unemployment is attributed is seeking behavior. Regarding a lack of process opportunities, Bruneians, however, are believed to be selective with modern-day method possibilities, with some qualified candidates claiming unsatisfactory pay, inconvenient walking hours, and adverse running conditions as reasons for rejecting task offers [4]. Furthermore, some are unwilling to take up temporary jobs, unconvinced that any type of revel received from such endeavors could assist their self-development while looking for the right mission [5]. Consequently, it can be concluded that unemployment among the younger Bruneians remains a growing concern. This is due to the lack of educational qualifications, limited job opportunities, and the selective and indifferent behavior of applicants. These factors have created additional problems for society, such as increased crime and deviance, as well as the emergence of runaway students who were initially expected to make a significant contribution to the country's development. It is crucial for the youth in Brunei to cultivate the right mindset and approach when considering job opportunities. They should strive to become more self-reliant and rely on their own efforts in job seeking to help address this issue in the country. If these attitudes are taught and nurtured by individuals, it is likely that by the year 2035, Brunei Darussalam will achieve its vision of establishing a dynamic and sustainable economy that provides its citizens with excellent employment opportunities for highly skilled individuals.

5.2.4 CONCEPTUAL FRAMEWORK

The conceptual framework is developed by the researcher after conducting a thorough literature review and aims to test the hypotheses related to the framework. It can be seen as the concrete concept or actual assessment of the researcher based on the purpose and objectives of the study. It is used to help understand the problem using a wide range of ideas and theories. The researcher gains a better understanding of the research problem and identifies the best solutions. Figure 5.6 illustrates the conceptual framework designed for this research. This framework can analyze the problems. In this case, the researcher wanted to analyze and discover the relationship between youth and the agricultural field in Brunei Darussalam in terms of perceptions, work availability, and the application of technological methods. This will work and provide a solution to the problem area as mentioned in the research question section above.

Within the framework, the researcher will focus on three variables: 'Youth,' 'Agricultural Field,' and 'Brunei Darussalam.' The researcher will examine the relationship between technology and agriculture, as well as seek to understand the processes of SMEs and businesses. Subsequently, the researcher will investigate the youth's perceptions of the agricultural field. Additionally, the researcher will explore the reasons for unemployment in

Brunei Darussalam, which may be related to the youth. This framework will assist the researcher in establishing connections between the variables and further elaborate on them in the discussion section.

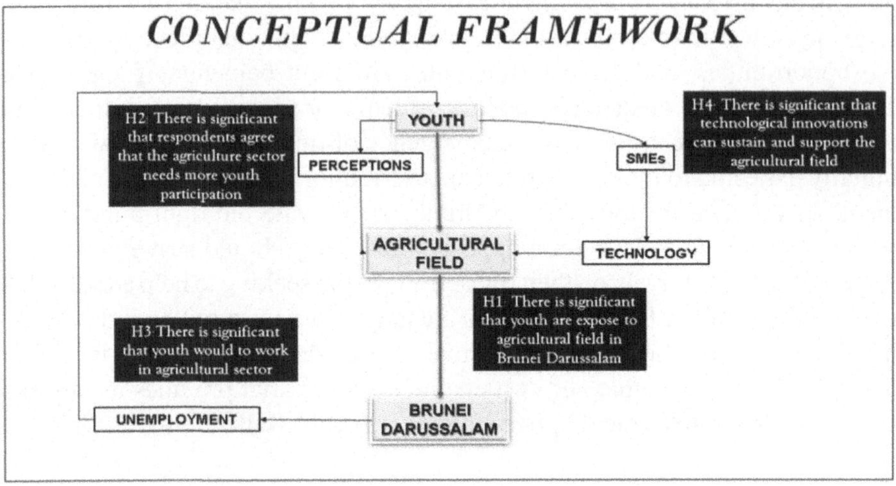

FIGURE 5.6 Conceptual framework.

The researcher will also correlate the literature reviews with the research questions and test the hypotheses. The study will examine the first theory, which investigates the extent of young people's involvement in farming in Brunei Darussalam. The second hypothesis aims to assess the respondents' perceptions regarding the need for increased youth engagement in the agricultural sector. The third hypothesis aims to determine the level of interest that youth may have in the agricultural sector. Moving on to the fourth hypothesis, the researcher intends to examine the potential of technological innovations to sustain and support the agricultural field. Table 5.1 presents the hypothesis test for the research study.

5.3 RESEARCH METHODOLOGY

5.3.1 RESEARCH DESIGN

For this study, the researcher will utilize a descriptive research approach. This study will employ a mixed research technique, incorporating both quantitative and qualitative methods, to understand the adolescent's perception of agriculture in Brunei Darussalam. The data for this study will be collected from both primary and secondary sources. The research will involve interviews, surveys to gather public opinions, and extensive review of scientific journals, papers, analysis notes, and policy or private sector research.

TABLE 5.1 The Hypothesis Test for the Research Study

Hypothesis I:

- **H_0:** There is no significant that youth are expose to agricultural field in Brunei Darussalam.
- **H_1:**There is significant that youth are expose to agricultural field in Brunei Darussalam.

Hypothesis II:

- **H_0:** There is no significant that respondents agree that the agriculture sector needs more youth participation.
- **H_2:** There is significant that respondents agree that the agriculture sector needs more youth participation.

Hypothesis III:

- **H_0:** There is no significant that youth would be interested to work in agricultural sector.
- **H_3:** There is significant that youth would be interested to work in agricultural sector.

Hypothesis IV:

- **H_0:** There is no significant that technological innovations can sustain and support the agricultural field.
- **H_4:** There is significant that technological innovations can sustain and support the agricultural field.

5.3.2 DATA COLLECTION

A qualitative systematic review can be conducted by reviewing all magazine articles related to research in Brunei's agriculture, specifically in the field of agriculture. Selecting the appropriate magazines will facilitate the researcher's task of reviewing existing studies on agriculture in Brunei Darussalam. The researcher will create a questionnaire using Google Form, consisting of both closed-ended and open-ended questions. Closed-ended questions can be answered with a simple one-word response, such as "yes" or "no." Closed-ended questions are typically found in formats such as multiple choice, drop-down, and ranking questions. In these cases, the researcher does not allow the respondent to provide spontaneous answers, but rather selects from a list of pre-selected options. These types of questions are designed to generate easily quantifiable data that can be coded efficiently, making them definitive in nature. They also enable researchers to categorize respondents into groups based on their selected options. On the other hand, an open-ended question is the opposite of a closed-ended question. Open-ended questions allow individuals to freely and spontaneously express their perceptions and opinions. It is designed to provide a comprehensive solution and generate detailed, qualitative data using the subject's personal knowledge and emotions. Open-ended questions also tend to be more exploratory and less restrictive than closed-ended questions. Subsequently, the survey will be distributed to gather public opinions on agriculture as a means of livelihood. The survey is designed in accordance with the research objectives. The questionnaires will be distributed to the general public through social media.

Due to the COVID-19 pandemic, researchers can successfully conduct interviews with selected SMEs in the agricultural field through online conventions. The interview questions may be in a semi-structured format and can be recorded using an audio recorder tool. The duration of each interview session should be at least 20 minutes and a maximum of 45 minutes. Researchers view that conducting an interview with a high-ranking officer is extremely important. This is not only to seek a precise perspective on this topic from the point of view of an insurance implementer, but also to gain a better understanding of the current situation of Brunei's agriculture. The findings from the interview can serve as evidence-based advocacy material for researchers in addressing this issue.

5.3.3 DATA ANALYSIS

For quantitative data, the researcher will compile the statistics using an Excel spreadsheet and SPSS software. Then, the researcher will analyze the data to answer each question. Excel can be used for data entry, manipulation, and presentation, as well as offering a range of statistical analysis capabilities and tools for running descriptive statistics and conducting various inferential statistical tests commonly used in business and management research. In addition, it offers all the identical antique spreadsheet functionality, which makes it useful for specific assessment and information manipulation tasks, along with generating graphical and wonderful presentation formats. Finally, even with the use of bespoke statistical software, Excel may be useful when preparing information for assessment within one's packages. Thus, Excel is an outstanding tool to use for assessment.

The SPSS software program utility is used for performing quantitative analysis and is a statistical package that is based on a point-and-click interface. This software has been widely utilized by researchers for quantitative analysis since its development in the 1960s by Norman H. Nie, in collaboration with C. Hadlai Hull and Dale Bent. SPSS is an abbreviation for Statistical Package for the Social Sciences, and it is used by researchers from various fields for complex statistical data analysis. The SPSS software program utility will automatically set up and import precise variable names, variable types, titles, and charge labels, meaning that minimum legwork is needed from researchers. Once survey data is exported to the SPSS software program, the opportunities for statistical analysis are almost endless. It will make the process of pulling, manipulating, and analyzing data easy and straightforward. This gives the researcher more time to do what they do best: identify trends, develop predictive models, and draw informed conclusions.

For qualitative data, the researcher will use critical questioning to synthesize the statistics. They will examine the interviews and utilize the data from the interviews to corroborate the findings with other data resources. Thematic assessment strives to identify styles of themes within the interview data. Thematic evaluation is one of the most popular methods for analyzing records. It focuses on a set of strategies that carefully examine the patterns that arise from the data collected from both primary and secondary sources. A benefit of using Thematic evaluation for analyzing data is that it is theoretically flexible. This evaluation method can be applied in various strategies and frameworks to address different types of research topics. Thematic evaluation can be particularly useful for studying people's opinions or perspectives from the participants' point of view.

5.4 FINDINGS AND RESULTS

As mentioned above, in this research, the data has been gathered using both primary and secondary records. In terms of secondary data, various references such as research journals, articles, study notes, and statistical reports have been utilized. On the other hand, for primary data, quantitative data was collected through a designed survey questionnaire consisting of 16 questions. The survey was designed based on the research objectives. This questionnaire employed random sampling, where the criteria for respondents were individuals who were willing to participate and answer the researcher's questionnaire.

5.4.1 SURVEY QUESTIONNAIRE

The questionnaires were then distributed to the public through various social media platforms, including Instagram, Telegram, WhatsApp, Facebook, Twitter, LinkedIn, and QR code. The survey questionnaires were primarily distributed through WhatsApp using a WhatsApp link and QR code, as shown in Figure 5.7. This platform was chosen for its convenience and accessibility, allowing potential respondents to quickly and easily access and answer the questionnaire. The survey data collection period lasted approximately three months. During the data collection, the researcher targeted only 150 respondents to participate in answering the questionnaire. However, the number of respondents exceeded the researcher's expectations, with a total of 208 recorded. The designed questionnaire consists of five different sections, including the background of the respondents, their employment and general opinions, their perceptions of work in the Agriculture field, and their opinions on youth and agriculture. The questionnaires generally contain five-point Likert scale questions, multiple-choice questions, open-ended questions, and closed-ended questions.

FIGURE 5.7 QR code to the survey link.

5.4.1.1 RESPONDENTS DEMOGRAPHIC

As shown on the pie chart in Figure 5.8, the majority of the respondents are female, accounting for 61%, while 39% of the respondents are male.

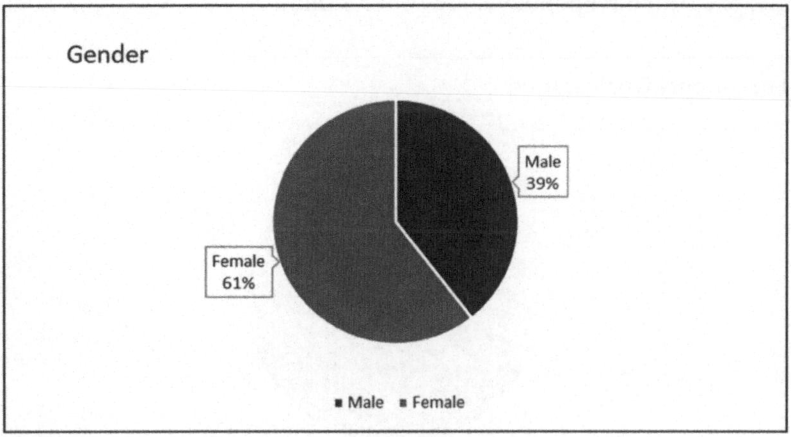

FIGURE 5.8 Respondents' gender.

Figure 5.9 shows the pie chart of the respondents' age groups. The age group that contributed the most to the survey is the category between 15 and 25 years old, with 50%, followed by the age group between 26 and 32 years old, with 30%.

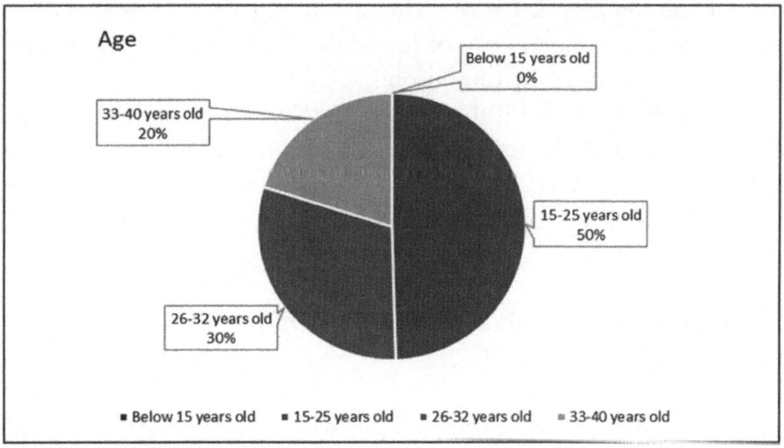

FIGURE 5.9 Respondents' group age.

In Figure 5.10, the Pie-Chart suggests the respondents' educational background. Approximately 45% of the respondents have an undergraduate academic qualification. This is followed by 24% of the respondents who have a post-graduate qualification. Around 15% of the respondents have a diploma, while approximately 10% have a school-level qualification. Additionally, 6% of the respondents are professionals.

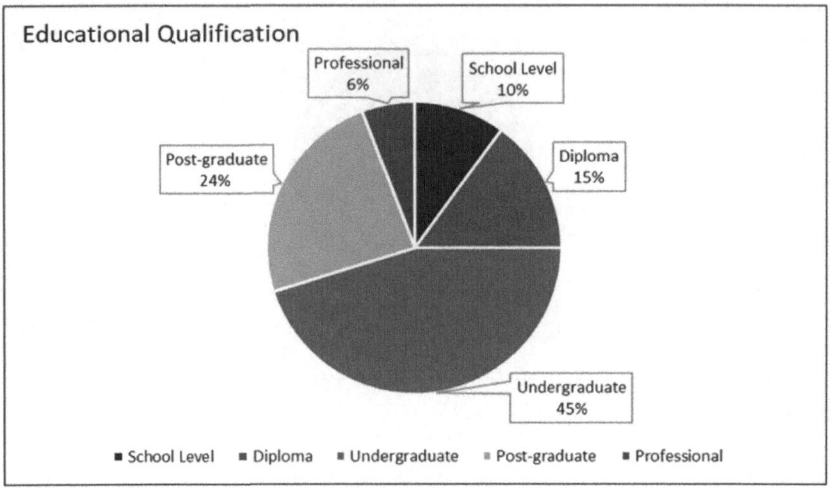

FIGURE 5.10 Respondents' education background.

The pie chart in Figure 5.11 indicates the respondents' institutional districts. The district with the highest number of respondents, accounting for 82% of the survey, is Brunei Muara. Tutong follows with 11% of the respondents. Approximately 6% of the respondents are from Belait, and 1% of the respondents are from Temburong.

As shown in Figure 5.12, the respondents' working status is as follows: 42% are employed, 41% are students, approximately 7% are unemployed, around 4% are part-timers, 3% are self-employed, and another 3% are retired. Table 5.2 provides a summary of the respondents' background.

5.4.1.2 RESPONDENTS AWARENESS ON THE OPPORTUNITIES IN AGRICULTURE FIELD IN BRUNEI

The researcher asked the respondents about their awareness of the opportunities associated with the agriculture sector in Brunei. The bar graph in Figure 5.13

illustrates the respondents' level of cognizance. Only 5% of the respondents are highly aware of the opportunities associated with the agriculture field in Brunei. The respondents come from various backgrounds including Food Science and Technology, Agriculture, Environmental Sciences, Food Security, and Food Economics. With this knowledge, they may understand how agriculture can provide opportunities in terms of employment, food security, sustainability, and economics.

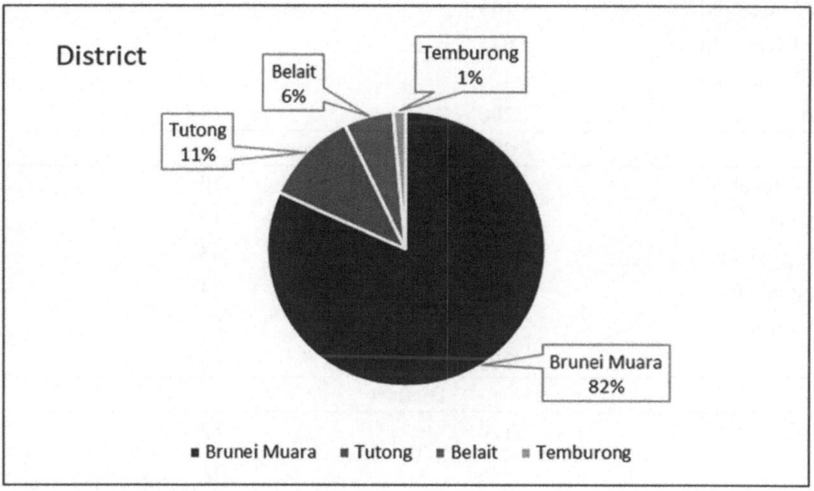

FIGURE 5.11 Respondents' group district.

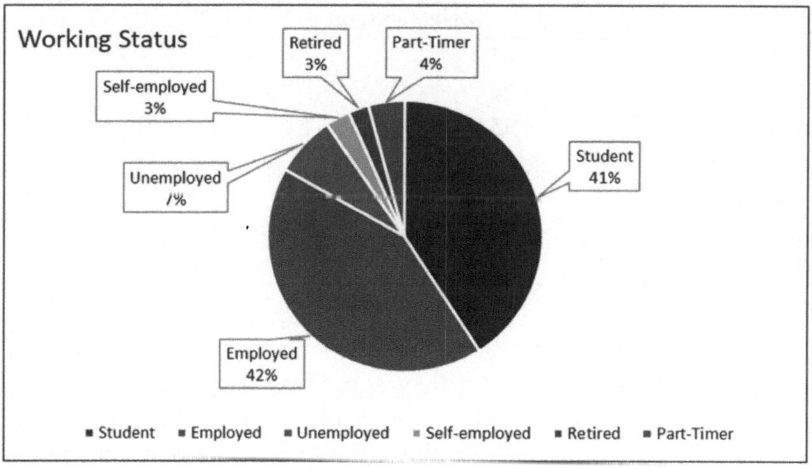

FIGURE 5.12 Respondents' working status.

TABLE 5.2 Summary of Respondent's Demographic

	Number of Respondents	Percentage (%)
	Gender	
Male	82	39
Female	126	61
Total	**208**	**100**
	Age	
Below 15 years old	0	0
15–25 years old	103	50
26–32 years old	63	30
33–40 years old	42	20
Total	**208**	**100**
	Education Qualification	
School level	21	10
Diploma	31	15
Undergraduate	94	45
Post-graduate	50	24
Professional	12	6
Total	**208**	**100**
	District	
Brunei Muara	170	82
Tutong	23	11
Belait	12	6
Temburong	3	1
Total	**208**	**100**
	Working Status	
Student	85	41
Employed	88	42
Unemployed	15	7
Self-employed	6	3
Retired	5	2
Part-timer	9	5
Total	**208**	**100**

The pie chart, as shown in Figure 5.14, depicts the responses of the respondents when asked whether they agreed or disagreed with the notion that the agriculture sector needs more youth participation. Approximately 98% of the respondents agreed that the agriculture sector does indeed require more youth participation.

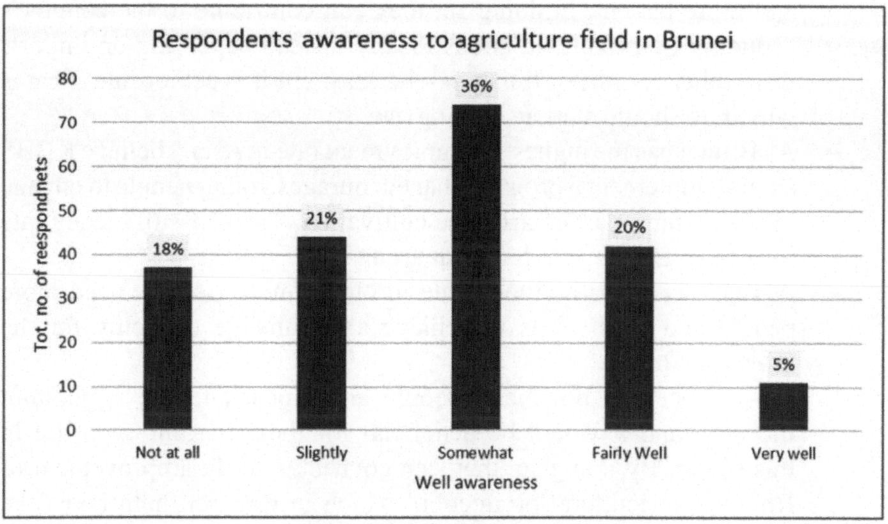

FIGURE 5.13 Respondents' awareness to agriculture field in Brunei.

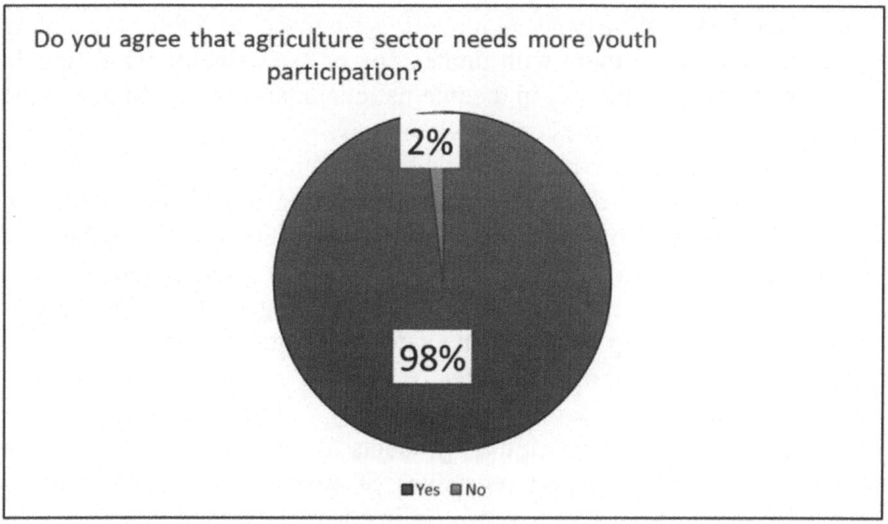

FIGURE 5.14 Respondents' opinion on agriculture sector needs more youth participation.

Various views and perceptions were received from the respondents regarding this statement. Some of the responses will be expanded as follows:

- Youth are the future generations. Through their participation, they can learn how to develop their beliefs and strengths to become better and more patient. In doing so, they can contribute to the country's economy by producing goods locally instead of relying on imports from other countries. It may not be easy, but it is achievable. That is why they should start at a young age.

- As Brunei has the highest unemployment rate in Asia, I believe KBBS should implement a program that encourages young people to engage in agriculture, particularly rice cultivation. The self-sufficiency rate for rice in Brunei is still low, at around 4%.
- Agriculture plays a crucial role in our economy, so we need more people and fresh ideas to achieve a sustainable economy for the Wawasan 2035.
- This is because not many people are exposed to the agriculture industry, and it would be beneficial for them to gain exposure in this sector. By doing so, they can contribute to the improvement of Brunei's agricultural production, which in turn can help boost the country's economy and reduce the unemployment rate. Additionally, pursuing a major in agriculture can also enhance skills and improve the overall quality of agricultural practices.
- Youth should be exposed more to agricultural technology, enabling us to produce more with limited space. Additionally, youth should explore opportunities in the international market to contribute to the improvement of the country's economy.
- The young generation is the backbone of our country. They need to be educated with a broader vision and a brighter future in the said sector. The young also have to be ambitious and determined in helping to shape the future of Brunei as an economically stable country."
- It is a great opportunity for young people to get involved in agriculture. It can help them become independent and not rely on working for the public or private sector.
- We cannot rely solely on the tertiary industry. Brunei does not produce enough agricultural products to support us Bruneians, and we still need to import vegetables. So why not? The youth needs to be aware of this and be proactive in changing their mindset.
- Nowadays, youth or the younger generation tends to focus solely on studying and may not be exposed to other fields. I believe it is important for young people to broaden their exposure, not only in this particular field but also in others.
- Young people's minds are very creative and innovative, which will result in the implementation of excellent products. They are the

right group to handle agriculture, as they are young, skillful, and energetic.

- From His Majesty's titah, the agriculture industry in Brunei needs to be further developed in order to boost local production and achieve self-sufficiency in terms of national food security.
- The youth need to take initiative in broadening the economy as per Wawasan 2035.
- Agriculture does not only involve growing and selling. It can also encompass market research and genetically modifying plants to adapt to Brunei's climate and requirements.
- Youths are the local people who will replace those currently working in the agriculture sector. By exposing them to the industry, some may develop an interest in studying it for various purposes, such as business or expanding their knowledge. This will lead to a greater variety of agricultural products.
- Because agricultural work requires the physical abilities of young people, as well as the open-mindedness and computer literacy of the new generation, which are beneficial for advancing in the commercial agricultural sector.
- There isn't much exposure to that field of knowledge among the youth. Yes, there may be some young people who are knowledgeable about the field, and some students who are learning about it. However, based on what I have seen through the news and other sources, there isn't much participation from people in my age group.
- So that the younger generation can assist the older generation in improving agriculture with new and improved technology. The younger generation also needs to learn from the older generation before it is too late, as in the future, they will be the ones responsible for ensuring the sustainability of Brunei's agriculture.
- The government has provided infrastructure and facilities in the agriculture/farming fields for locals to participate in producing agricultural products. This is to ensure food security and a continuous food supply in the country.
- Because the involvement of youth is still too low in this field, we need them to utilize their knowledge and talent in technology to participate in the agriculture sector for the future.
- As part of Brunei's 2025 vision to transition into the non-energy sector, the government should invest in agricultural studies for higher education instead of prioritizing students to work in the oil and gas industry.

- Who will continue the legacy if not our youth? They need to immerse themselves more in this field to acquire knowledge and experience, rather than relying solely on the government sector.
- Youth can be a catalyst for getting involved and participating in the agricultural sectors, as they make up a high percentage of the whole population in Brunei.
- The basics of sustainability are, without a doubt, farming. The implementation of IoT is common, as shown by aggrotech. Since today's youth are exposed to technologies, they are well-suited for this.
- Conventional farming may involve basic labor. I have observed, even as a student, while working part-time and volunteering, that many young people (not all) are picky or under the delusion that obtaining a degree or high qualification guarantees employment. This has been my personal experience, and I hope the situation improves in some way.
- We need the youth to continue the progress of agricultural efforts and become successors to the older generation, rather than relying solely on them.
- This has created a balance in terms of workforce supply and demand. Additionally, there is currently a lack of supply in primary resources.
- To impart knowledge about the agricultural sector to the future generation and inspire innovation among them.
- The agriculture industry is vital for human survival and will always have demand.
- I believe the agriculture sector is not a popular career choice among Bruneians, especially the youth.
- Yes, because youth participation will generate more fresh ideas to help in the development of the agriculture sector. It will also help the government achieve their vision of GenerasiBerwawasan 2035, which emphasizes the youth's contribution to the country.
- The government needs fresh ideas that could be provided by the youth, as well as a means to reduce the unemployment rate in Brunei.
- This is because our youth are capable of learning new things, and it is also due to the country's goals in 2035. Why not consider training our youth under certain conditions and contracts? I believe they would be willing to seize this opportunity.
- Yes, with the close guidance from those experts in the field, youths who studied agriculture at IBTE should be taught and trained by farmers/experts in farming for the hands-on part, while IBTE focuses

on the theoretical part. This is the best synergy, rather than relying solely on IBTE's in-house lecturers who may be fresh graduates or have little experience in agriculture.

- Yes, with the close guidance from those experts in the field, youths who studied agriculture at IBTE should be taught and trained by farmers/experts in farming for the hands-on part, while IBTE focuses on the theoretical part. This is the best synergy, rather than relying solely on IBTE's in-house lecturers who may be fresh graduates or have little experience in agriculture.
- Agriculture has the potential to be even bigger than it is now, serving as a significant source of economic growth for the country. Therefore, young people should be more open to this opportunity and anticipate the industry's growth, which offers numerous job opportunities.
- It is not only for more job opportunities, but more importantly for the sustainability of the country.
- The youth may change the agricultural sector for the better by finding more ways to increase efficiency, pouring more research studies, and making modifications.
- Because the youth represent the future of the nation, and they are the most dependable group of citizens, their energy and determination would suffice for developing a successful agricultural organization.
- Since we are living in a challenging and more competitive world, it is now the time to generate more innovative ideas for the younger population.

The pie chart, as shown in Figure 5.15, represents the responses of the respondents when asked if they would be interested in working in the Agriculture Field. Approximately 75% of the respondents expressed their interest in working in the Agriculture Field, while 25% indicated that they are no longer interested.

Various explanations were obtained for this argument from the respondents. Many of the factors that the respondents may like to operate in the area of agriculture are as follows:

- Not only will it be a good experience for me, but it will also have the chance to boost the Bruneian economy, apart from depending on our oil and gas. We are too dependent on oil and gas. Once those run out, Brunei will face its worst economic crisis. I would like to contribute to this sector towards a more sustainable future.

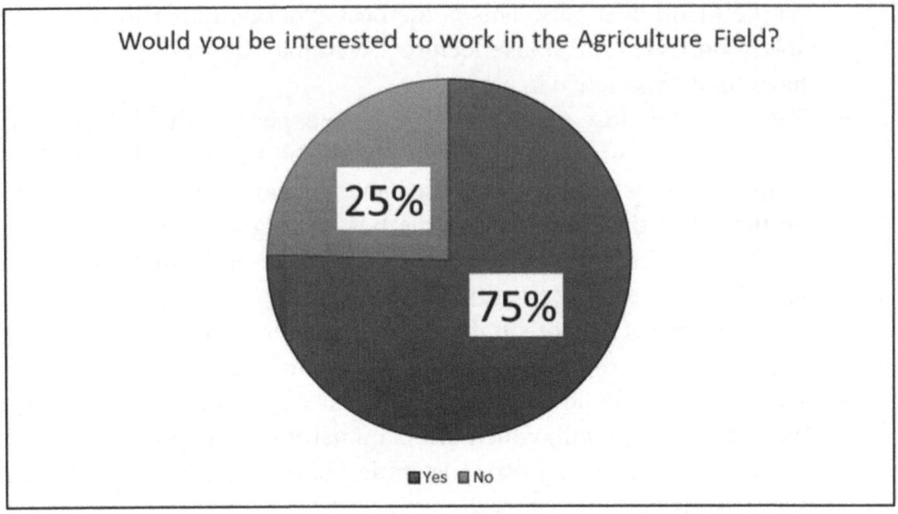

FIGURE 5.15 Perception of respondents regarding working in the agriculture field.

- The agricultural scope of the field has not been normalized; therefore, there might be a lot of room for new knowledge to be gained. I believe that Brunei has the potential to increase its economy and sustainability through agriculture. However, we do have limitations such as a lack of skilled labor and limited space.
- Sense of achievement comes from knowing that what you did can have a positive impact on the economy, and it is a way to give back to the country.
- It would be an interesting field to work in, considering the land and weather in Brunei. There is potential for a variety of plants to be cultivated and tested in the laboratory.
- I would like to work in this field as it is related to my studies in the Food and Beverage industry. This sector has the potential to improve F&B businesses by reducing the cost of agricultural products, which are primarily used in the industry.
- With exposure and guidance, I would like to start my own business in agriculture to provide food for both the local and foreign markets.
- To ensure food security and maintain a consistent and adequate food supply, it is important to reduce dependence on foreign products, especially in the event of disruptions or loss of food supply chains from outside Brunei. Additionally, diversifying Brunei's economy beyond reliance on oil and gas is crucial.

- Diversifying skills and labor, and not relying solely on government and private sector employment, agriculture is a business that can contribute to sustaining the economy.
- Agriculture is essential for any country. This is what you call a sector of daily necessity. If we decrease competition and all can work under one big community, to invest and to avoid competition, the agriculture sector can grow at an effective rate and experience growth.
- As I mentioned before, it is important for us to carry on the legacy. I would also like to gain more knowledge and learn new things by working in the field of agriculture, alongside experienced farmers.
- Agriculture serves as the foundation of human societies. It is a sector that grows exponentially with the population. As such, it always offers a very stable line of work, as its demand will always increase.
- It not only benefits ourselves but also our country in the future, without relying too much on the government. By doing this, we can help our government and our country, as well as future generations. Doing good for our country is also doing good deeds. We can also engage in business independently, just like other countries, which helps to boost our economy.
- Something different. Agriculture has been highlighted many times as one of the crucial areas that will support the future economy, so I would like to contribute my part to this.
- Agriculture can be exciting given the right opportunities and initiatives.
- I would like to give this a try, as it can help increase the availability of local food resources for the public. Additionally, it could serve as an alternative financial source for the country, alongside oil.

5.4.1.3 PERCEPTIONS OF RESPONDENTS' WORK IN THE AGRICULTURE FIELD

Figure 5.16 shows the pie chart of the respondents' preferred types of farming. The most preferred is Commercial farming, practiced for the purpose of creating a profit, with 22%. This is followed by Fish farming, also known as Aquaculture, with 16%. Approximately 14% of respondents opt for Mixed farming, which involves the development of vegetation and rearing of animals on the same portions of land. Around 10% of respondents choose Arable farming, which involves the cultivation of crops in warm climates. Another

10% of respondents prefer Subsistence farming, the practice of growing plants and rearing animals for personal use. About 8% of respondents opt for Extensive and Intensive farming, which includes land tillage with the aim of increasing output. Next, 5% of respondents choose Sedentary farming, which entails tilling an equal area of land for many years. The least favorable options are Pastoral farming, the practice of rearing animals only in cold and moist climates, not suitable for growing crops, and Nomadic farming, which involves farmers moving with their animals from one location to another in search of pasture and water, with 2%.

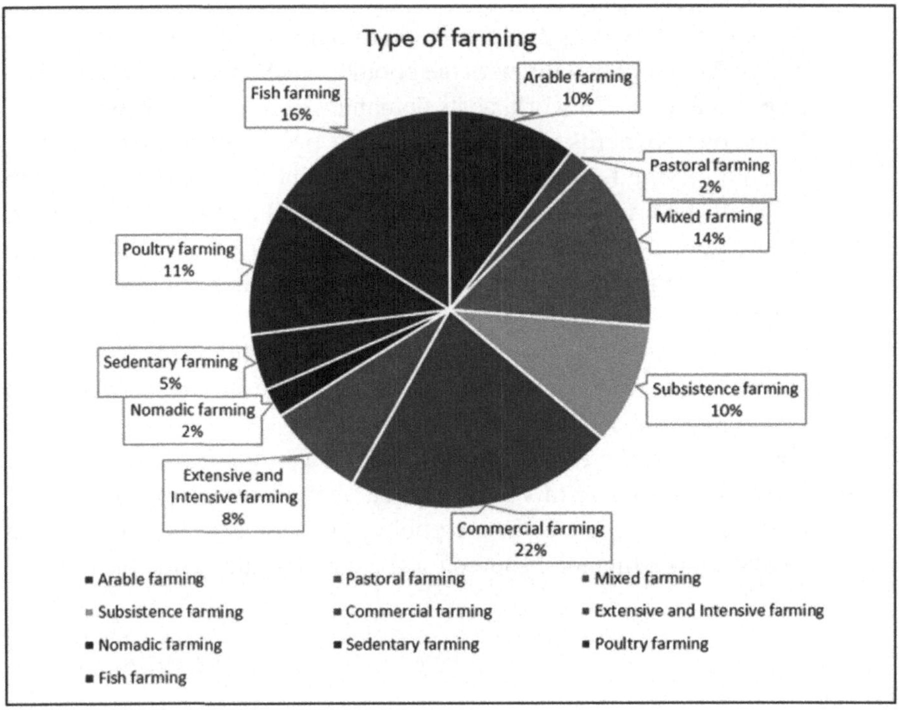

FIGURE 5.16 Type of farming, preferably.

The pie chart in Figure 5.17 shows the respondents' preferred method of farming. Almost all respondents prefer both traditional and modern methods of farming, with 74%. Approximately 25% of respondents prefer the modern method of farming, while only 1% of respondents still opt for the traditional method of farming.

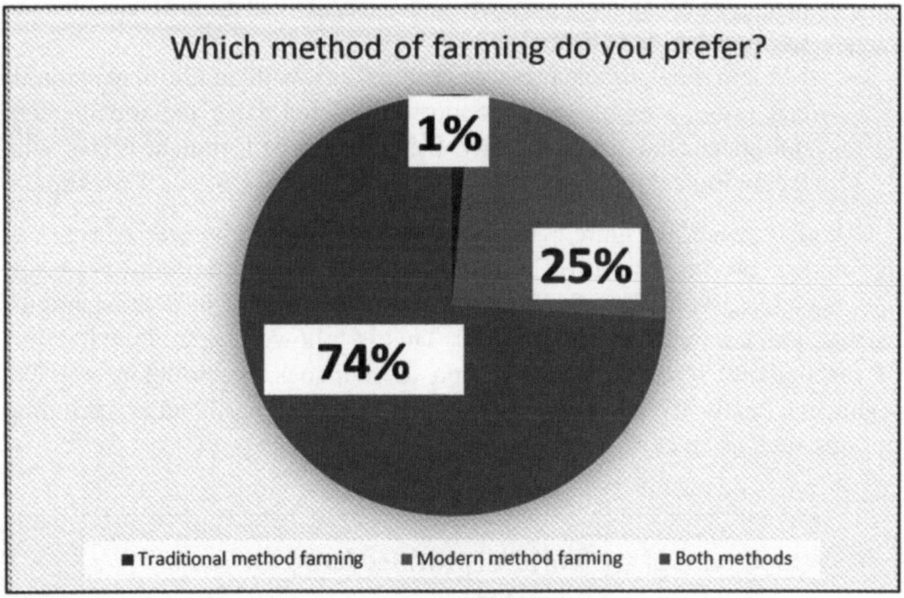

FIGURE 5.17 Respondents have a preference for farming methods.

There are specific explanations provided by respondents on both conventional and modern farming practices, and some of the answers were as follows:

- There are ways that are both easy and hard to maintain. One can be found in either method. So, it is better to know both so one can find alternatives in growing goods.
- Some modern methods may be more efficient, but using traditional methods can also be useful at times.
- Using both methods could help to improve farming methods. The traditional way can be more effective as people are able to practice it hands-on, while the modern way utilizes technology to make farming faster and more advanced.
- Because traditional methods are usually practical, while modern ones are convenient. Sometimes, the old method will be more efficient and cost-saving than the new modern technology that can yield enough to make a profit. However, integrating older methods and modern methods when necessary is always best.
- Incorporating traditional methods, when necessary, can help reduce costs compared to using modern ones. However, modern methods

cannot be completely neglected because they can definitely contribute to faster yields at times.

➢ Modern methods can increase productivity, while traditional methods can provide employment opportunities for those lacking in technological skills. Additionally, in the event of technical issues with technological methods, we can rely on manual work as a backup.

The bar graph in Figure 5.18 shows the respondents' preferred types of agricultural produce or crops. The top three most preferred produce or crops are vegetables, with 17% of respondents selecting options such as asparagus, carrots, lettuce, tomatoes, onions, etc. This is followed by fruits, with 14% of respondents selecting options such as mangoes, watermelon, papaya, logan, etc. Lastly, livestock accounts for about 12% of preferences, including options such as chicken, cattle, sheep, goat, and duck.

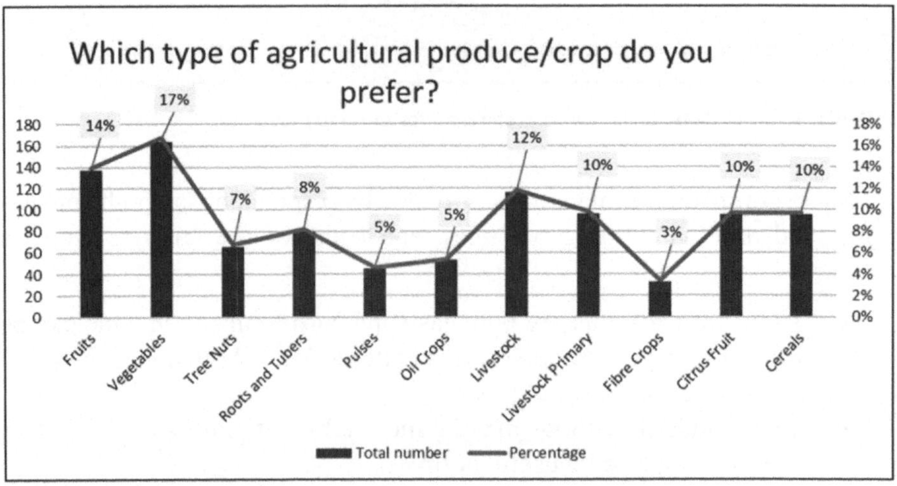

FIGURE 5.18 Respondent's preferred type of agricultural produce or crop.

5.4.2 INTERVIEW

5.4.2.1 SMALL MEDIUM ENTERPRISE (SME) WITH AGRICULTURE RELATED

The researcher also managed to interview several parties from private companies involved in the agriculture field. These selected companies include BFB Agrotechnology Company, S&R Aquafarm, and Kemari Farm

Enterprise. They willingly participated and shared their perspectives and experiences on the issues related to the agriculture field. Their inputs greatly helped the researcher analyze and further understand youth exposure in the agriculture field, which will be elaborated on in the further discussion section. However, during the tough time of the COVID-19 pandemic, the researcher re-adjusted the data collection strategy. Alternative methods that were used included online video conferencing platforms such as Zoom Online Meeting and Skype. Table 5.3 provides a summary timetable of the interview sessions with the selected SMEs involved in the study case.

TABLE 5.3 The Summary Timetable of the Interview Sessions

Company	Online Video Conference/Site Visit	Time	Day and Date
1. S&R Aquafarm	Site visit (Address: No. 162, Kg Penanjong, Daerah Tutong)	9.00 am–10.30 am	Sunday, 26 April 2020
2. BFB Agrotechnology Company	Via Zoom	9.30 am–10.30 am	Tuesday, 12 May 2020
3. Kemari Farm Enterprise	Via Zoom	9.30 am–10.30 am	Tuesday, 5 May 2020

5.4.2.2 BFB AGROTECHNOLOGY COMPANY

Iffah Billah binti Haji Junaidi, the director of BFB Agrotechnology company, a Brunei-based Aquaponics system, hopes to position Brunei as the first Solar City in the foreseeable future. Their business enterprise aims to reduce carbon footprints and elevate solar-powered technology to new heights in Brunei. to encourage households to cultivate their own food, a Bruneian agrotechnology organization is prepared to implement a simple, affordable, and accessible farming method.

According to her, aquaponics refers to the combination of plants grown in water and aquatic species, such as fish and prawns. This makes growing food easier, as one does not need to worry about soil or fertilizers. The goal is to simplify the farming process, allowing people to choose the size of their farms. People can grow fruits or vegetables indoors, in outdoor ponds, in an aquarium, or even on their desks by growing flowers. Families can also rely on their aquaponic garden for food during emergencies or natural disasters. By encouraging more families to grow their own food, the company partners with government agencies to achieve self-sustainability in the country. The company also aims to educate more children, particularly young people interested in farming, and encourage them to start their own gardens from a young age. The children need to be taught early on about the importance of food sustainability and how they can contribute to it. An aquaponic garden would be a suitable way for them to learn, as it requires minimal maintenance. Aquaponics not only allows for the growth of plants, but also provides an opportunity to raise fish or prawns in the ponds or aquariums. Fish and prawns are well-suited for aquaponics because they don't rely on soil for plant growth. The waste produced by the fish or prawns can serve as fertilizer, providing the necessary nutrients for the plants.

During the interview, she focuses on who will continue the legacy if not our youth. They need to be exposed more to this field to gain knowledge and experience, rather than relying solely on the government sector. The youth represent the future, and agriculture helps to diversify the economy and ensure survival, rather than solely depending on government jobs. They should be exposed to the agriculture field at an early stage so that the younger generation can help the older generation improve agriculture with new and improved technology. The younger generation also needs to learn from the older generation before it is too late, as they will be responsible for ensuring the sustainability of Brunei's agriculture in the future. Without the youth, we would rely solely on the older generation, and that is not sustainable. Youths are the local people who will replace those currently working in the agriculture sector. By exposing them to the industry, some may develop an interest in studying it further, leading to business opportunities and expanding knowledge in agriculture, resulting in a wider range of agricultural products. This is important because not many people are exposed to the agriculture sector, and it would be beneficial for them to explore it, as it could help improve Brunei's production and economy, as well as reduce unemployment rates. Additionally, specializing in agriculture can enhance skills and the quality of agricultural practices. It is suggested to introduce agriculture as a subject at the primary level, as this is when children begin to explore new concepts. This will help them develop an appreciation for those working in this sector. Furthermore, organizing farming classes under CCA for youths and the public would provide valuable experience and generate interest. Make it interesting and stimulating for them. Also, emphasize that although agricultural work can be physically demanding, the hard work and dedication can lead to rewarding outcomes. Promote this message to the younger generation, particularly middle school-aged groups, and highlight the use of modern smart technology in agriculture to improve productivity and catch their attention, fostering curiosity.

5.4.2.3 S&R AQUA FARM

SyazwanHj Suni, the founder of S&R Aqua farm set up aquaponics business with his father in 2011 in their backyard in Tutong and had bagged the most innovative Enterprise and third place in Sustainable Business. A Brunei-based family-owned business aim to educate, share, and spread the awareness of fresh, organic produce and free from chemical pesticide. There is an almost intuitive appeal for aquaponics. At its core, aquaponics is a system that cultivates fish, typically in a tank, while simultaneously utilizing their waste to grow hydroponic vegetables. The fish excrements are collected and pumped into a separate tank where the vegetables' roots are submerged in water. The waste, rich in ammonia, attracts bacteria that convert it into beneficial nitrates, serving as a natural fertilizer for the plants. Additionally, the plants act as a filter, purifying the water before it is returned to the fish tank. This closed-loop system mimics the symbiotic relationship found in natural habitats like lakes and rivers, abundant with vegetation and aquatic life [6].

During the interview, Syazwan mentioned that our youth are capable of learning new things, and this is also due to the country's goals in 2035. Why not consider training our youth under certain conditions and contracts? I believe they would be willing to seize this opportunity. The government needs fresh ideas that can be contributed by the youth in order to reduce the unemployment rate in Brunei. Syazwan emphasizes that youth participation will generate more innovative ideas to aid in the development of the agriculture sector. This will also help the government achieve their vision of *GenerasiBerwawasan 2035,* which emphasizes the youth's contribution to the country. With the support of senior employees in the agriculture sector, there will be higher chances for success in the agriculture sector for Brunei. He added that youth can be a catalyst to get involved and start participating in agricultural sectors since they make up a high percentage of the whole population in Brunei. Based on experience, he stated that agriculture is not a risky sector to venture into and there are opportunities for business in agriculture. He recommended implementing technology in farming methods, as well as introducing fresh ideas to the industry, to attract the youths and utilize their capabilities in improving the agricultural sector.

5.4.2.4 KEMARI FARM ENTERPRISE

Kemari Farm is a start-up farming company run by young agro-entrepreneurs, aiming to provide an incubation program that teaches essential farming skills to create job opportunities for youth. This farming start-up enterprise, operated by young individuals, aims to offer teenagers an incubation program that imparts crucial farming competencies, ultimately leading to the creation of job opportunities. Recently, a group of five young people who committed to training 50 Bruneians to become farm owners through an incubation program won the best social business concept and received $10,000 from Standard Chartered Brunei. This achievement came as part of the fourth Youth Against Poverty initiative organized by the Society for Community Outreach and Training [7]. Kemari Farm has already been given a 0.7-hectare plot at the Salambigar Youth Development Center after showing interest in the Young Farmers Program at the center. Co-founder Muhd Hafizuddin Hj Awg Damit said Kemari will sustain itself by commercially growing crops, with local youth learning the entire farming cycle of different crops from Kemari's farmers.

During the interview, the co-founder agreed that the agricultural field needs more youth participation. He mentioned that agriculture has the potential to be bigger than it is now, such as being a source of economy for the country. Therefore, youths should be more open to this opportunity and participate more, as this industry offers a lot of job opportunities. The youth may change the agricultural sector for the better by finding more ways to increase efficiency, through conducting more research, studies, and making modifications. He explained that the reason why the youth represent the future of the nation is because they are the most dependable group of citizens. Their energy and determination would suffice for developing a successful agricultural organization. He added that since we are now living in a challenging and more competitive world, it is high time to generate more innovative ideas for the younger population. He advised that hands-on learning is important for many youths. In addition to being able to more fully understand educational and agricultural concepts, youth will have fun completing agricultural activities. A booklet should be prepared that provides step-by-step instructions and a list of materials needed for the activity to make coordinating educational activities fun and simple.

5.4.3 STATISTICAL ANALYSIS

5.4.3.1 ONE-SAMPLE T-TEST ANALYSIS

The one-sample t-test analysis is used to test the four hypotheses mentioned above. The one-sample t-test is a part of the t-test family. All tests in the t-test family compare differences in mean continuous-level, categorical or ratio scores, which are normally distributed. The one-sample t-test operates on a single mean score, unlike the independent or dependent-sample t-test. The one-sample t-test compares a single sample mean to a fixed value to assess if the mean of the study is significantly greater or less than that value. '0' represents No and '1' represents Yes. The decision rule for assessing if the test is significant for $\alpha = 0.05$ is as follows: if $p \leq 0.05$, the study is considered significant, while if $p \geq 0.05$, the study is not considered significant. Figure 5.19 shows the formula for the one-sample t-test.

$$t = \frac{\bar{x} - \mu}{\frac{s}{\sqrt{n}}}$$

FIGURE 5.19 One-sample t-test formula.

For the first hypothesis, H1, it refers to identifying whether the youth are exposed to the agricultural field in Brunei Darussalam. Table 5.4 is the result from the One-sample t-test, which shows that a p-value of less than 0.05 implies significance, and that of less than 0.01 implies high significance. Therefore, $p = 0.0000$ implies high significance. High significance means that the statement of evidence is true in order to reject the null hypothesis. As the p-value is less than 0.05, the test is significant, and the sample is significantly different than $\mu = 0$. The youth are exposed to the agricultural field in Brunei Darussalam $t\ (207) = -11.49$, $p = 0.000$. At the 95% confidence interval of the difference, ranging from –0.46 to –0.32, it does not include zero. So clearly, we accept H1, which means that there is a relationship between youth awareness and agriculture.

For the second hypothesis, H_2, the aim is to determine whether the agriculture sector requires more youth participation. Based on the results of the One-sample t-test, as presented in Table 5.5, the Agriculture sector shows a significant need for more youth participation, $t\ (206) = -2.02$, $p = 0.045$. The p-value is less than 0.05. However, since the 95% confidence interval of the difference ranges from –0.46 to 0.00, which includes zero, we reject H_2. This implies that there is no significant evidence to support the claim that agriculture needs more youth involvement.

TABLE 5.4 One-Sample Test for Hypothesis H_1

One-Sample Statistics

	N	Mean	Std. Deviation	Std. Error Mean
Are you aware and expose of the opportunities related to agriculture field in Brunei	208	0.61	0.489	0.034

One-Sample Test

	Test Value = 1					
	t	df	Sig. (2-tailed)	Mean Difference	95% Confidence Interval of the Difference	
					Lower	Upper
Are you aware and expose of the opportunities related to agriculture field in Brunei?	-11.490	207	0.000	-0.389	-0.46	-0.32

TABLE 5.5 One-Sample Test for Hypothesis H_2

One-Sample Statistics

	N	Mean	Std. Deviation	Std. Error Mean
Do you agree that agriculture sector needs more youth participation?	207	0.98	0.138	0.010

One-Sample Test

	Test Value = 1					
	t	df	Sig. (2-tailed)	Mean Difference	95% Confidence Interval of the Difference	
					Lower	Upper
Do you agree that agriculture sector needs more youth participation?	-2.015	206	0.045	-0.019	-0.04	0.00

For the third hypothesis, H_3, the objective is to determine whether any of the respondents would be interested in working in the field of Agriculture. According to the results of the One-sample t-test, as shown in Table 5.6, a p-value of less than 0.05 indicates significance, while a p-value of less than 0.01 indicates high significance. Therefore, a p-value of 0.0000 indicates high significance. As stated in the first hypothesis, high significance implies that the evidence supports the rejection of the null hypothesis. Since the p-value is less than 0.05, the test is considered significant, and the sample is significantly different from $\mu = 0$. There is a significant indication that respondents would be interested in working in the agriculture field if given the opportunity, with t(206) = -8.21, p = 0.000. Since the 95% confidence interval of the difference ranges from -0.31 to -0.19, it does not include zero. Therefore, we clearly accept H_3, which implies a connection between respondents and their interest in working in the agriculture field.

TABLE 5.6 One-Sample Test for Hypothesis H_3

One-Sample Statistics

	N	Mean	Std. Deviation	Std. Error Mean
If you have the opportunity, would you be interested to work in the Agriculture Field?	207	0.75	0.432	0.030

One-Sample Test

	Test Value = 1					
	t	**df**	**Sig. (2-tailed)**	**Mean Difference**	**95% Confidence Interval of the Difference**	
					Lower	**Upper**
If you have the opportunity, would you be interested to work in the Agriculture Field?	-8.206	206	0.000	-0.246	-0.31	-0.19

For the fourth hypothesis, H_4, it is recognized that technological innovations can sustain and support the agricultural field. According to the One-sample t-test result, as shown in Figure 5.7, t (207) = –1.74, p = 0.083. The p-value is greater than 0.05. Since the 95% confidence interval of the difference ranges from –0.03 to 0.00, which includes zero, we reject H_4. This means that there is no relationship between technological innovations support and the agricultural field (Table 5.7).

TABLE 5.7 One-Sample Test for Hypothesis H_4

One-Sample Statistics

	N	Mean	Std. Deviation	Std. Error Mean
Do you agree that technological innovations can sustain and support the agricultural field?	208	0.99	0.120	0.008

One-Sample Test

	Test Value = 1					
	t	df	Sig. (2-tailed)	Mean Difference	95% Confidence Interval of the Difference	
					Lower	Upper
Do you agree that technological innovations can sustain and support the agricultural field?	-1.740	207	0.083	-0.014	-0.03	-0.00

5.4.3.2 DESCRIPTIVE STATISTICS ANALYSIS

Descriptive statistics is a timeframe provided for statistical evaluation that allows us to describe, display, or summarize facts in a meaningful way. This helps us identify patterns or trends from the information. On the other hand, qualitative statistics does not enable us to draw conclusions about the knowledge we have reviewed or make assumptions. The researcher's goal was to explore technological innovations that can support and enhance the agriculture field. A five-factor scale was used, where '0' indicates strongly disagree, '1' indicates disagree, '2' indicates neutral, '3' indicates agree, and '4' suggests strongly agree. Based on the results, as shown in Table 5.8, the range of scores for 'suggest' is between 3.44 and 3.57. The highest mean rating is 3.57, suggesting that making agriculture more profitable is important, followed by a mean rating of 3.65 for improving agriculture's image. The lowest mean score is 3.44, which suggests linking social media to agriculture.

Table 5.9 shows that all the histogram results for the technological innovations that can sustain and assist the agriculture field.

TABLE 5.8 Descriptive Statistics Analysis Results

Descriptive Statistics					
	N	**Minimum**	**Maximum**	**Mean**	**Std. Deviation**
Link social media to agriculture	207	0	4	3.44	0.721
Improve agriculture's image	207	0	4	3.56	0.657
Strengthen higher education in agriculture	207	0	4	3.49	0.696
Greater use of information and communication technologies (ICT)	207	0	4	3.52	0.667
Empower young people to speak up	207	0	4	3.40	0.774
Facilitate access to land and credit	207	0	4	3.40	0.730
Put agriculture on the school curricula	207	0	4	3.46	0.733
Greater public investment in agriculture	207	0	4	3.47	0.681
Make agriculture more profitable	207	0	4	3.57	0.671
Valid M (listwise)					

5.5 FURTHER DISCUSSION

5.5.1 THE INCREASING OF UNEMPLOYMENT RATE IN BRUNEI

The pie chart in Figure 5.20 illustrates the opinions of the respondents regarding the rise in unemployment in Brunei. Approximately 39% of the respondents identified a mismatch of skills in the job market as a contributing factor. This issue can be further exacerbated by the challenges of acquiring new skills that are specific to emerging markets and technological advancements, making it difficult for unemployed farmers to secure employment in high-tech industries. Moreover, finding suitable housing or schools for their children can also pose challenges during the job search. Technological transition is another factor, as the introduction of labor-saving technology in certain sectors can lead to a reduced demand for the types of labor that have been replaced by machinery. Around 24% of the respondents indicated a preference for remaining unemployed rather than accepting available jobs. If the benefits are favorable, individuals may choose to continue receiving benefits instead of seeking employment. Frictional unemployment is often a form of voluntary unemployment, as individuals prefer to wait until a better job opportunity arises. Approximately 14% of respondents mentioned

TABLE 5.9 Histograms Results

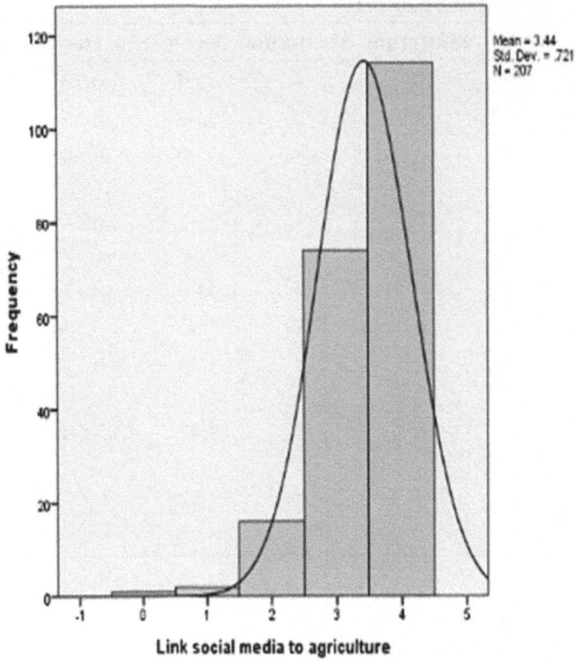	The distribution of the link social media to agriculture is negatively skewed distribution. 't' has a peak to the right of center, greater step by step tapering to the left side. It is linear, with the right mode and greater than either average or average. The mean is nearer to the left and lower than the mean and median. This shape suggests that every outlier is less preponderant than normal. The mean is 3.44 and standard deviation is 0.721.
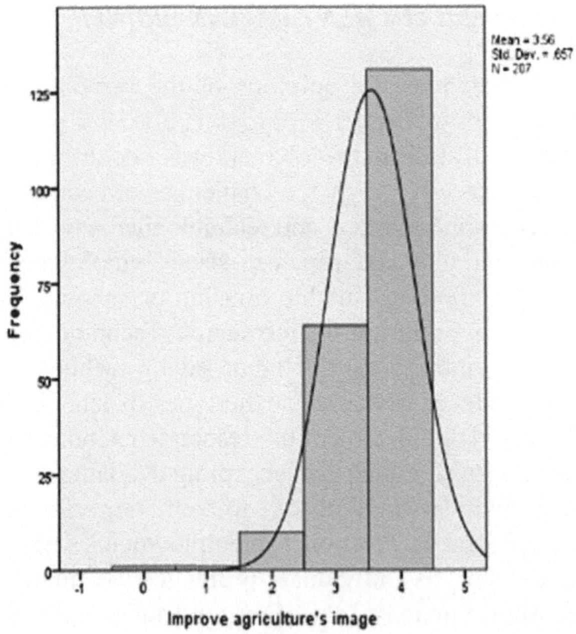	Distribution of raising the reputation of agriculture is negative biased distribution. It has a point to the right of center, tapering more slowly to the left. With mode nearer to the right and greater than either median or average, it is symmetric. The mean is 3.56 and standard deviation is 0.657.

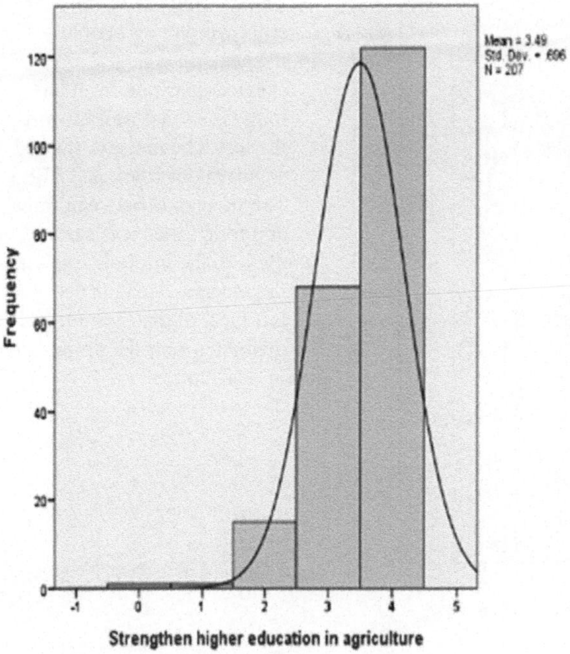

The distribution of the strengthen higher education in agriculture is negatively skewed distribution. It has long tailed that trail towards the left. The mean is 3.49 and standard deviation is 0.696. The median is between three and four. Since it is skewed to the left, the mean is much less than the median. Mean near the left aspect of distribution and median is close to the proper aspect of distribution.

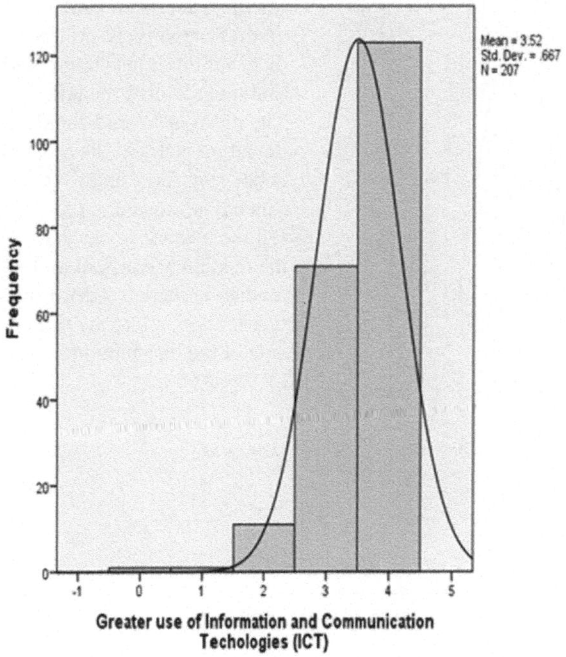

The distribution of the greater use of information and communication technologies (ICT) is negatively skewed distribution. It has long tailed that trail towards the left. The mean is 3.52 and standard deviation is 0.667. The median is three or four. Since it is skewed to the left the mean is lower than the median. Mean near and median to the left facet of the distribution is much like the right facet of the distribution.

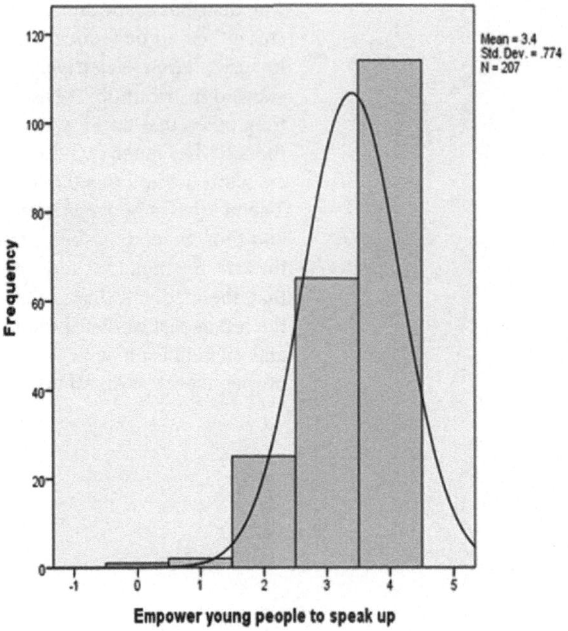

The distribution of the empower young people to speak up is negatively skewed distribution. It has long tailed that trail towards the left. The mean is 3.4 and standard deviation is 0.774. The median is between three and four. Since it is skewed to the left, the imply is less than the median. Mean is near the left facet of distribution and median is near the proper side of distribution.

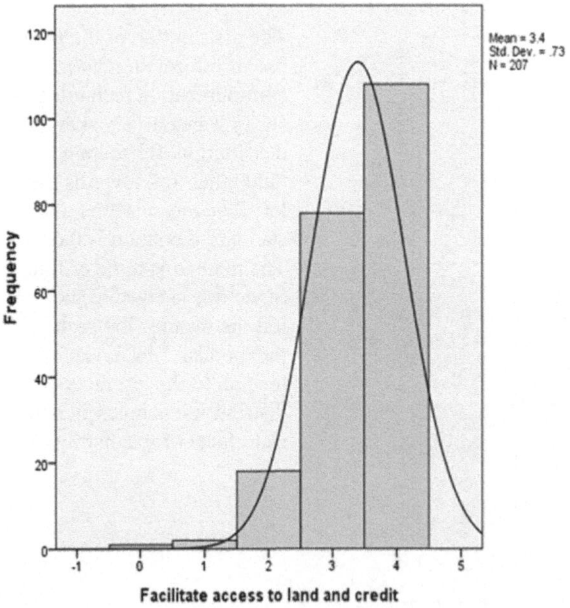

The distribution of the facilitate access to land and credit is negatively skewed distribution. It has long tailed that trail towards the left. The mean is 3.4 and standard deviation is 0.73. The median is between three and four. Since it is skewed to the left the mean is lower than the median.Mean near and median to the left side of the distribution is like the right side of the distribution.

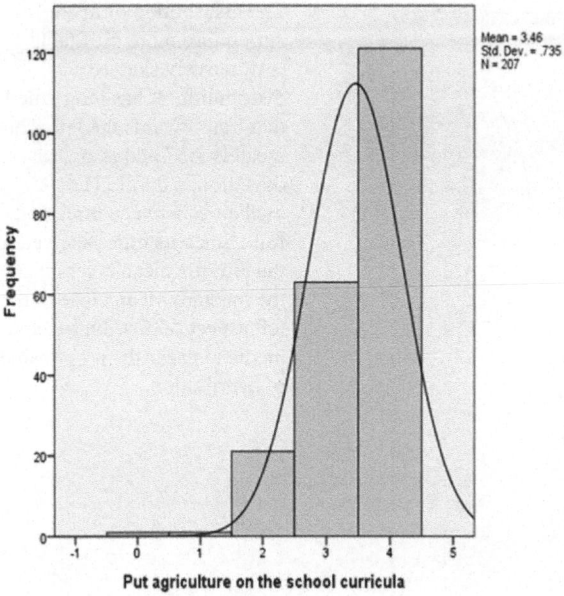

The distribution of the put agriculture on the school curricula is negatively skewed distribution. It has long tailed that trail towards the left. The mean is 3.46 and standard deviation is 0.735. The median is between three and four. Since it is skewed to the left, the suggest is much less than the median. Mean close to the left aspect of distribution and median is close to the right aspect of distribution.

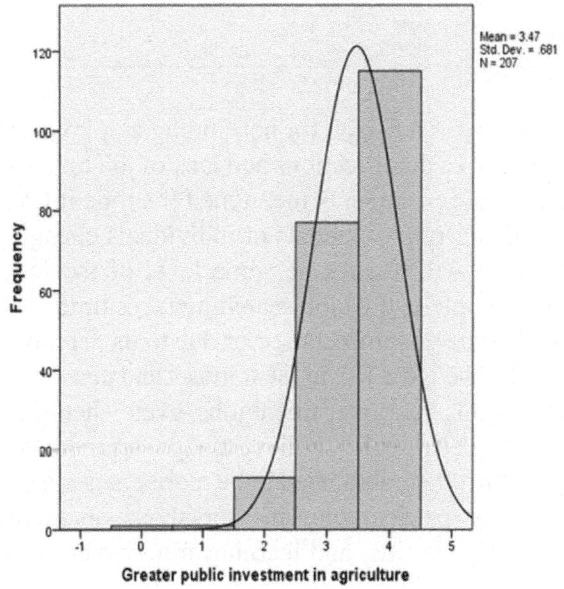

The distribution of the greater public investment in agriculture is negatively skewed distribution. It has long tailed that trail towards the left. The mean is 3.47 and standard deviation is 0.681. The median is between three and four. Because it is skewed to the left the mean is lower than the median. Mean close to the distribution side and median is close to the distribution side.

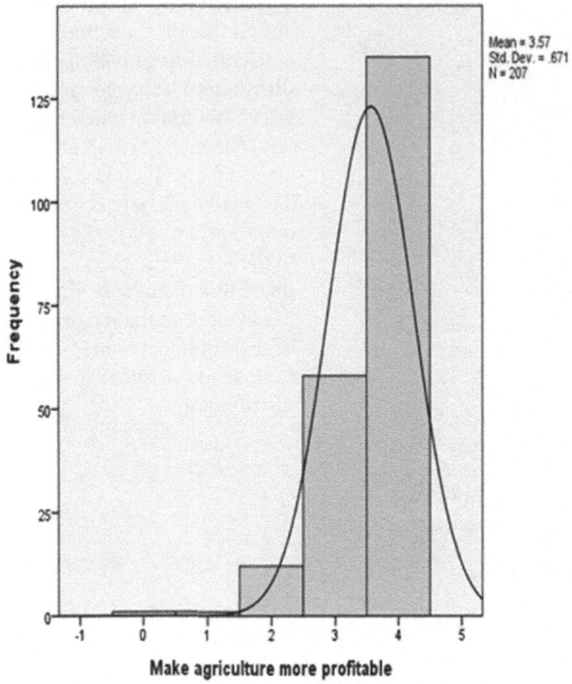

The distribution of make agriculture more profitable is negatively skewed distribution. It has long tailed that trail towards the left. The mean is 3.57 and standard deviation is 0.671. The median is between three and four. Since its miles skewed to the left, the mean is less than the median. Mean close to the left aspect of distribution and median is near the proper side of distribution.

insufficient demand in the economy as a reason for not finding employment, followed by 13% who cited excessive qualifications and lack of job opportunities. Additionally, around 10% of respondents mentioned the time it takes to transition between jobs, particularly for students or individuals changing careers. Within an industry, there will always be some level of frictional insecurity as knowledge is not complete, thus job searching takes time.

Most people prefer to work in the government sector due to its perceived security and higher pay. However, the issue lies in the mindset and perception of the youth. They tend to apply only for government jobs, even when there are vacancies in the private sector, because they believe that government jobs offer stable income. To address this, the youth need to be more open-minded and exposed to opportunities by participating in workshops, attending seminars, joining entrepreneurial programs, and learning from experienced individuals.

Other than that, the lack of interest among unemployed individuals may stem from their perception that they should be working in a field related to their degree, rather than taking on temporary jobs such as bartending or waiting tables. One problem with unemployed youths is their lack of discipline, such as not attending interviews on time. There is also a need for better alignment between academia and industry, as current industrial needs are not being adequately identified. To address this, Brunei's educational institutes, including universities and technical institutes, should focus more on producing graduates for local industries rather than prioritizing global rankings. It is necessary to shift from the traditional approach to more hands-on and industry-ready programs, as continuing with the current norm would result in graduates who are less marketable for local industries. The Brunei government needs to conduct a critical review and analysis of the ethos and methodology, for example, that was previously adopted and practiced by the Institute Teknologi Brunei (ITB) in their Higher National Diploma (HND) programs. During those years, ITB HND graduates were highly marketable and favored by local industries, making them industry-ready. The HND programs were also widely accepted for their high standards and quality by universities in the United Kingdom (UK) and around the world, thus gaining global recognition and acceptance.

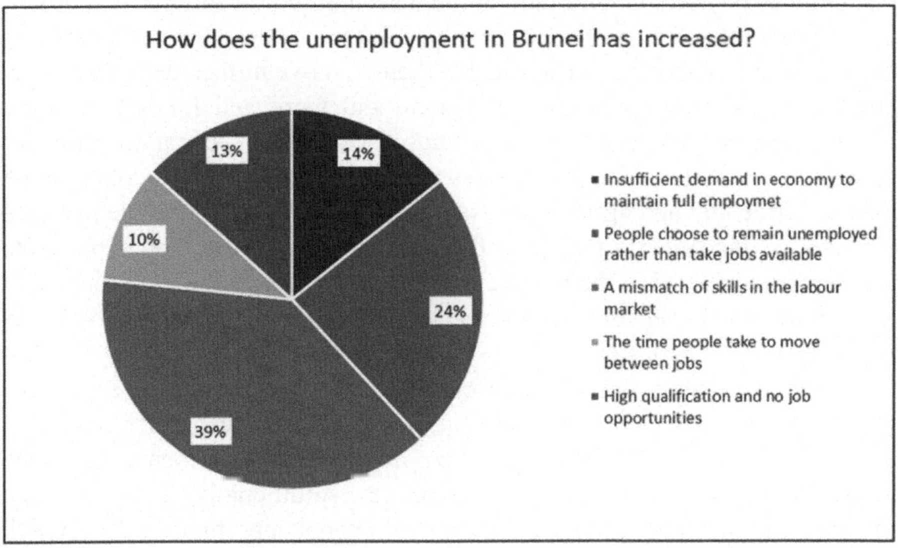

FIGURE 5.20 Opinions of respondents on the increase in unemployment in Brunei.

Moreover, employers tend to retain their staff by extending their contracts rather than hiring new people. Additionally, most jobs require prior experience, which poses a challenge for fresh graduates who lack experience. There is a higher demand for jobs with good qualifications than the number of available positions in the market. This leaves many graduates unable to find suitable employment, as the number of graduates continues to increase each year.

5.5.2 THE CAUSES OF DECLINE IN YOUTH PARTICIPATION IN AGRICULTURAL WORK

According to the framework, the researcher states that it is significant for the youth in Brunẹi Darussalam to be exposed to the agricultural field. In terms of psychological factors, this consideration involves certain inherent values, expectations, or perceptions about agriculture and the circumstances surrounding the program's implementation. In this context, the psychological state of mind of the youth is a crucial factor due to the conditions that influenced the initiative and their long-term aspiration for white-collar jobs. This is attributed to the disconnection from farming, preference for quick economic activities, and the perception of farming as outdated and low-status occupations. Perhaps due to modernization, young people tend to avoid agricultural work that requires physical labor. Many of them have little exposure to the industry, which leads to a lack of interest. As a result, their knowledge and interest in the agricultural sector diminish, causing them to move further away from rural areas. Negative perceptions about agricultural work and limited exposure during childhood contribute to this mindset. Nowadays, education primarily focuses on technology and emphasizes the benefits of working indoors, rather than on gardening and agriculture. Additionally, some young people perceive farming as unprofitable and lack the passion for it. Most youth prefer to work in fields related to their qualifications, and if they do have an interest in agriculture, it is better for them to start gaining relevant education rather than studying in a different field.

In terms of environmental factors, there are several considerations that primarily affect the youth who are willing to participate but are unable to do so due to distance, program duration, and inadequate infrastructure such as power, internet, and recreational facilities. Additionally, there is a lack of fertile and acidic soil, as well as limited export opportunities. The youth often perceive agricultural work as physically demanding, conducted in harsh environments, and offering low wages. This perception may be influenced by the availability of off-farm job opportunities, as well as the processes of urbanization and globalization. Furthermore, the agricultural department itself lacks in terms of marketing efforts. One of the reasons for this is the failure to promote the various types of agricultural work to the younger generation, who are more inclined towards technology and may not find this field of work interesting. The reason for the decline in local agriculture is the lack of demand for local products. The only way for local farmers to sell their produce is by selling it themselves. They are unable to sell to supermarkets because supermarkets in Brunei import food from outside the country and have their own farms. Local farmers cannot compete with this. Additionally, agricultural jobs are physically demanding and not preferred by many. The hot sun and long hours make it challenging, and the results are not immediate. Furthermore, farmers are only paid when they sell their products. In today's era, technology plays a significant role in our lives. Due to a lack of awareness about modern agricultural practices, some people may think that farming still relies solely on traditional and time-consuming methods. This misconception discourages young people from participating in agriculture.

In terms of government-induced factors, there have been failures to fully implement the empowerment program, limited engagement, dual reporting, and inadequate communication. Additionally, the government has not made any interesting agricultural facts readily available to the public. There is also a lack of initiatives that promote fun activities or advertisements to encourage youth to learn about agriculture.

5.5.3 ACTIONS CAN BE DONE BY THE YOUTH TO SUPPORT AGRICULTURAL WORK

In the research framework, it is stated that there is a significant need for more youth participation in the agricultural sector. Agriculture is one of the rare sectors where you do not need an agriculture degree or qualifications to get started. The youth need to change their mindset. Thus, youth involvement can improve Brunei's economy. Actions that can be taken by the youth include showing interest in this area. However, this can only be achieved if their elders and the government demonstrate how interesting this field of study is. They need to be more involved, active, and participate in this area. They need to be exposed more in order to help build their interest in the field of agriculture. It is important for them to be more aware of the "survival kit" based on agricultural work, as this can change their mindset and show them how interesting it is to work in the agricultural sector and how to make a profit from it. There should be more public events focused on agriculture to encourage participation. This will help people understand how exciting this job can be. While many young people focus on innovative farming methods, it is equally important to consider the next step, such as finding innovative ways to sell their produce.

Both the government and non-government sectors play crucial roles in promoting agriculture. They should organize farming awareness workshops, seminars, and training sessions to highlight the importance of agriculture. Additionally, they should focus on developing technology to enhance agricultural practices. It is essential to raise awareness about the benefits of entering the agriculture sector, emphasizing its profitability and sustainability due to the consistent demand. The agricultural department, in particular, should engage in extensive marketing targeting the youth, such as sponsored advertisements on social media. Offering paid internships and practical work experiences in agricultural areas can provide firsthand understanding of the farming process to the youth. Furthermore, they can promote the opportunities for success and high salaries with job security in the agricultural field. Conduct group events or activities within the agriculture industry to provide early exposure to the youth. Promote the various work scopes to the youth, aiming to increase their knowledge and generate interest in working in the sector. This initial exposure may spark their interest. Provide more opportunities for suitable activities that will engage the youth and capture their interest. Introduce them to different aspects of agriculture, highlighting that it is not solely about manual labor. Emphasize the importance of gathering and analyzing data for the growth of plants or products, as well as making informed business decisions.

In terms of education, strengthening higher education in agriculture could provide numerous benefits. It is crucial to introduce agricultural courses at an early age. Additionally, organizing school trips and farm visits can effectively educate and raise awareness about the importance of agricultural education. Encouraging youth participation in agricultural activities is essential. Offering Agricultural Science, Food Security, and Sustainability as major subjects in higher education and universities for undergraduate programs would be beneficial. Moreover, agriculture subjects should be made available to students with a strong foundation in Mathematics, Sciences (Chemistry and Biology), and Business. The aim is to promote these subjects to younger generations, especially middle school age groups, and to raise their interest in modern smart technology used to enhance agricultural productivity. To capture their attention and foster curiosity, a dedicated program should be established. This program will attract, engage, and retain them in the field of agriculture, enabling them to become experts and successful entrepreneurs in agricultural activities for their livelihood.

Essentially, the youth need to be trained from an early stage. Their interest can be increased by discussing successful individuals in the agriculture sector. Campaigns and other activities should be created to attract the youth. It is especially important to provide education on agriculture at the primary, secondary, high school, and institute levels. Implementing an annual agricultural-focused activity in every school can help change the mindset of the youth. Recognizing their achievements and offering lower payment as an incentive is crucial. Support is crucial as youth are motivated by it. They require guidance and early mentorship to build confidence and generate interesting ideas for agriculture. These actions can contribute to the research framework, indicating a significant likelihood of youth working in the agricultural sector.

5.5.4 TECHNIQUES TO ATTRACT YOUTH TO JOIN THE AGRICULTURE FIELD

Based on the framework, the researcher discovered the respondents' opinions on the level of agreement regarding techniques to involve youth in agriculture. In Figure 5.21, the researcher asked the respondents if linking social media to agriculture is considered an effective technique to attract youth to engage in the field. About 189 respondents agreed with this statement. The rise of social media and its popularity among young people, combined with access to appropriate technologies, could pave the way for agriculture. The use of mobile smartphones is rapidly growing in other countries, connecting people to information and each other. Utilizing these platforms to promote farming and educate young individuals could greatly impact new groups of people in the field. This has the potential to increase youth interest and engagement.

In Figure 5.22, improving agriculture's image is another technique to attract youth to join the agriculture field. Approximately 196 respondents agreed with this statement. Farming is rarely portrayed in the media as a sport for younger individuals and is often seen as outdated, unprofitable, and challenging work. There is a need to raise awareness among young people about the benefits of agriculture as a profession, particularly the opportunities for increased consumer participation, creativity, and innovation within the farming sector. The media, ICT, and social media can all be utilized to enhance the perception of agriculture among a wider audience and facilitate the exchange of knowledge and experiences between young individuals and farmers. By making agriculture more interesting and innovative, we can encourage youth to be more receptive and engaged in the field.

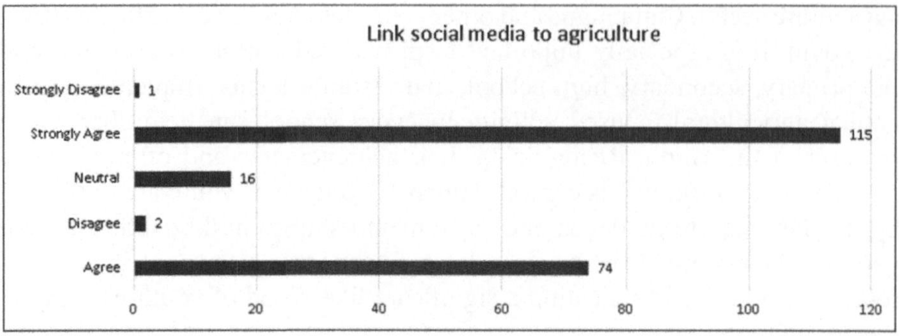

FIGURE 5.21 Link social media to agriculture.

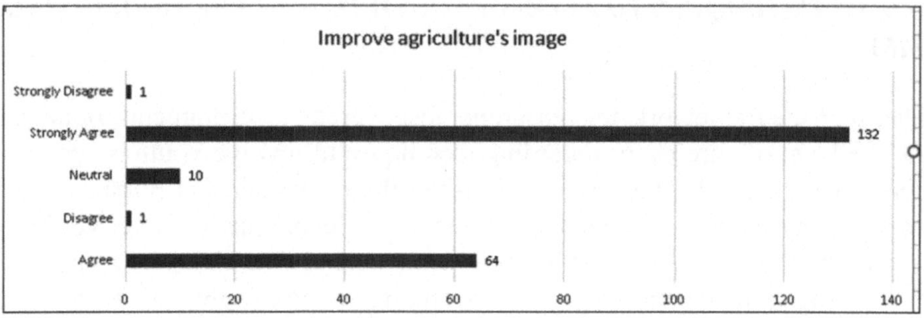

FIGURE 5.22 Improve agriculture's image.

In Figure 5.23, approximately 191 respondents agreed to strengthen higher education in agriculture. Relatively few students choose to pursue agriculture, possibly due to the mixed quality of farm education. The materials taught need to be connected to technological advancements in order to promote innovation and be more relevant to a dynamic and evolving agricultural market, with a focus on agribusiness and industry. In addition to technical skills, developing management, decision-making, communication, and leadership abilities should also be emphasized in higher education. Agricultural tertiary schooling reforms should be focused on young individuals, involving them directly in the process. Strengthening higher education from an early stage will undoubtedly benefit the youth in considering their future.

In Figure 5.24, a greater use of Information and Communication Technologies (ICT) is also another technique to attract the youth to join the field. Approximately 195 respondents agreed with this statement. Not only can ICT be used to educate and prepare those who are unable to access higher education institutions, but it can also serve as a platform to facilitate knowledge sharing, create networks, and pursue employment opportunities for young individuals. Technological solutions will be necessary to cater to a technologically savvy generation. These innovations can also enhance market transactions, thereby increasing agricultural productivity. In this new era, technology has been widely used in various aspects, necessitating a change in farming methods to improve overall efficiency.

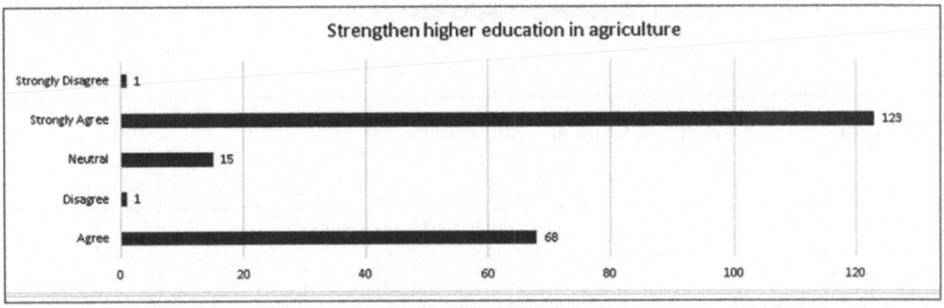

FIGURE 5.23 Strengthening higher education in agriculture.

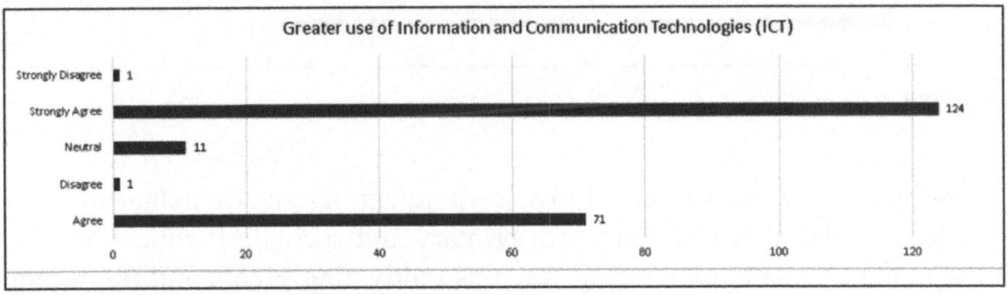

FIGURE 5.24 Greater utilization of information and communication technologies (ICT).

In Figure 5.25, approximately 180 respondents agreed to empower younger people to speak up. Brunei should provide a platform for young people to discuss possibilities in agricultural development, share success stories, and advocate for increased youth engagement and representation. Giving the youth a chance to speak up and listening to them would undoubtedly bring great potential to the country in the agricultural field.

In Figure 5.26, approximately 187 respondents have agreed that facilitating access to land and credit can attract the youth. Property is always expensive and difficult for young people to purchase, making it nearly impossible to obtain loans for property acquisition without leverage. Innovative investment is required for agriculture and small businesses. As a starting point, providing soft loans to young individuals who come up with novel farming or micro franchising ideas.

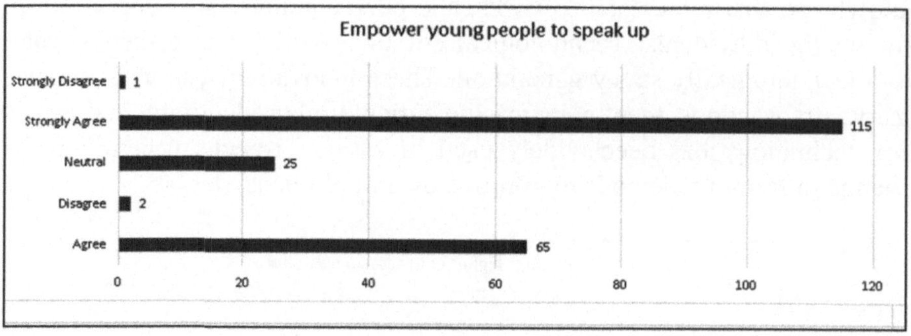

FIGURE 5.25 Empower young people to speak up.

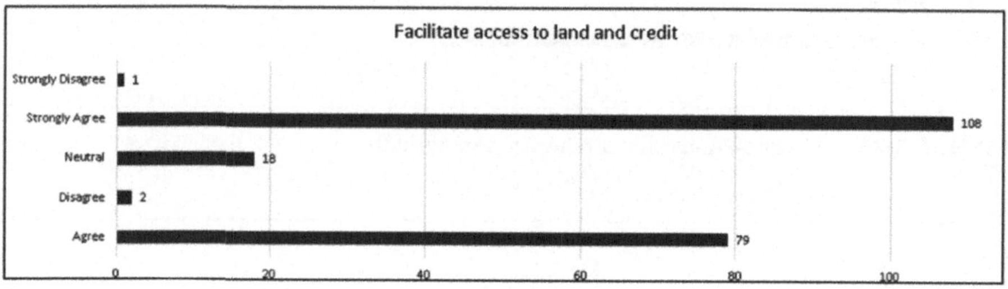

FIGURE 5.26 Facilitate access to land and credit.

In Figure 5.27, approximately 185 respondents agreed to include agriculture in the school curriculum. Both primary and secondary education may incorporate elements of farming, such as cultivating plants and promoting crops. This could inspire young individuals to consider farming as a potential career.

In Figure 5.28, approximately 193 respondents have agreed to provide extra public funding in agriculture. Young people may perceive agriculture as a sector that the government has largely neglected, giving farmers the image of being old-fashioned. Agricultural investment is more effective in reducing poverty than investment in any other sector, yet public spending on agriculture remains low.

In Figure 5.29, approximately 194 respondents agreed that making agriculture more profitable is important. Ensuring the sustainability of agriculture involves reducing the costs of cultivation and doing business, while simultaneously increasing production. While large-scale commercial farming may come to mind, this is not always the case, as small farms can be highly productive with low labor costs.

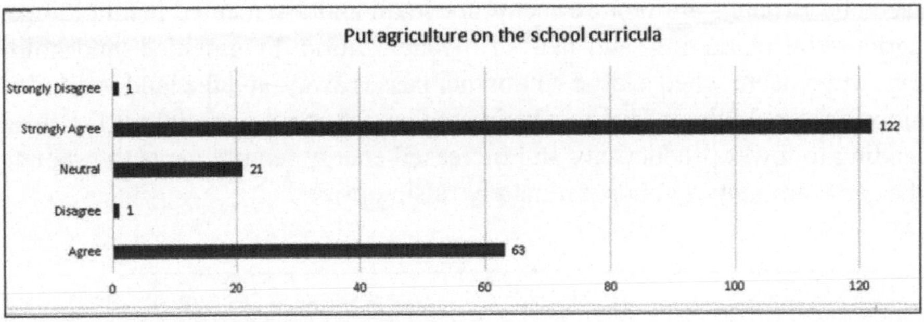

FIGURE 5.27 Put agriculture on the school curriculum.

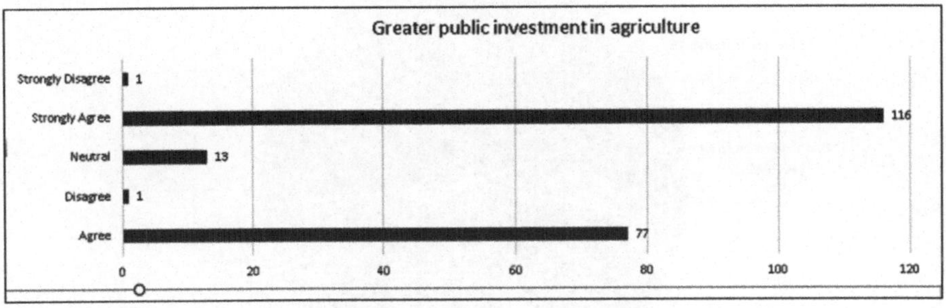

FIGURE 5.28 Greater public investment in agriculture is needed.

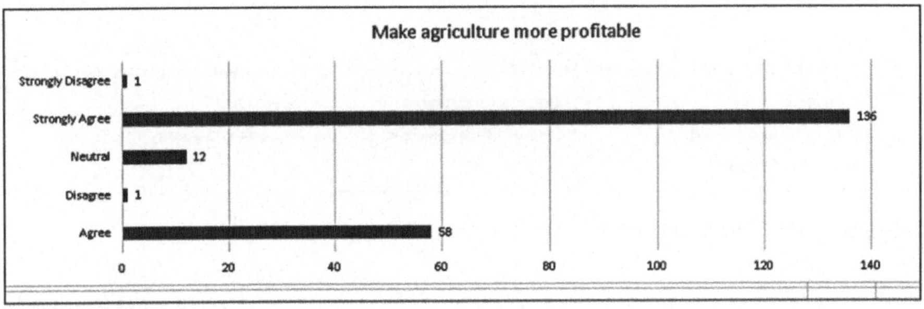

FIGURE 5.29 Make agriculture more profitable.

5.5.5 THE PROBLEMS BRUNEI ENCOUNTERS IN TERMS OF AGRICULTURE

The pie chart in Figure 5.30 illustrates the most common problems and issues encountered by Brunei in terms of agriculture. The top three challenges faced by Brunei's agriculture sector are small and fragmented landholdings, agricultural marketing, and lack of mechanization. Fragmented land holdings often occur when a large landowner passes away, and the land is divided among their children. This results in smaller farms for agricultural families, leading to lower productivity and increased energy requirements to produce the same amount of goods, ultimately raising prices.

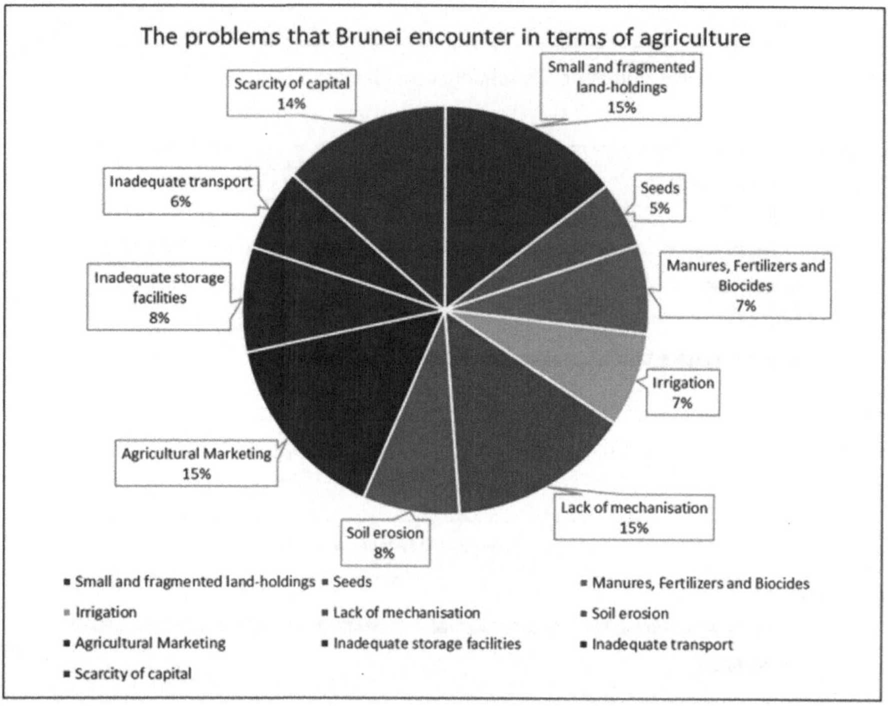

FIGURE 5.30 The problems that Brunei encounters in terms of agriculture.

Agricultural marketing encompasses various aspects of the business environment and program, both practical and structural, that revolve around technological and economic factors. This includes pre- and post-harvest activities such as assembly, processing, transportation, and delivery. However, Brunei still faces challenges in terms of agricultural advertising. These challenges include a low marketable surplus of agricultural goods, lack of control over pricing by producers, storage issues, transportation problems, a lengthy chain of middlemen, malpractices within the marketplace, lack of market information, inelastic demand, and absence of grading systems.

These factors contribute to the cumbersome nature of agricultural marketing in Brunei.

Lack of mechanization in agriculture is another problem in Brunei. Rising levels of mechanization no longer necessarily mean large investments in tractors and other machinery. Farmers need to choose the most suitable power source for each operation, depending on the work to be done and who is doing it. Mechanization in general will easily and reliably meet their needs. Sustainable agricultural mechanization can also have a significant impact on the development of supply chains and food markets, as it can make post-harvest, production, and distribution processes and operations more effective, efficient, and environmentally sustainable.

Different explanations were obtained from the respondents regarding the problems Brunei faces in relation to agriculture. Many of the explanations provided by the respondents were as follows:

- *"The agricultural scene is not being taken seriously by the government. Therefore, it is impossible to set aside profits for maintaining this field."*
- *"Brunei has only recently started diversifying into agriculture. More investments towards agriculture are needed."*
- *"If you want to start up, you might need a high cost, especially to acquire the land and machines required."*
- *"Most of the fruits are imported from another country."*
- *"The country needs to improve its agricultural marketing so that it can attract customer attention and provide more information about itself."*
- *"Most of the seeds are imported, which will affect Brunei's sustainability. The farming sites are far from water reservoirs, therefore requiring transportation to transport water or the use of underground water, thus disrupting the land. There is a lack of skilled labor and limited machinery, techniques, and land, resulting in the reliance on traditional methods and slow progress."*
- *"Brunei's geographical reasons for farming (acidic soil, etc.) and government red tape."*
- *"Incapable of purchasing, replacing, or repairing and updating with the latest amenities."*
- *"The government should prioritize making agriculture more appealing and rewarding to the younger generation."*
- *"Agricultural marketing is of utmost importance because Brunei does not lack in any of the issues mentioned above, except for agricultural marketing. Even in terms of agricultural marketing, it is not lacking in my opinion. What is lacking is the demand for local produce. The*

only solution is to not only sell their produce here, but also outside of Brunei."

- *"The best support products (e.g., fertilizer) are not available in Brunei Darussalam, so we need to purchase them from another country."*
- *"Not a straightforward investment."*
- *"Because the soil in Brunei is predominantly acidic, it can hinder the growth of crops. As a result, the cost of farming is high due to the frequent need to maintain soil fertility by adding chemical substances to reduce acidity."*

5.5.6 TECHNOLOGICAL INNOVATIONS THAT CAN SUSTAIN AND SUPPORT THE AGRICULTURE

Based on the conceptual framework, implementing technology in farming methods and introducing fresh ideas to the industry could sustain and support agriculture. The youth have the capability to contribute fresh ideas to improve the agricultural sector. According to the pie chart shown in Figure 5.31, the researcher asked the respondents whether they agreed or disagreed that technological innovations can sustain and support the agricultural field. Approximately 99% of the respondents agreed with the statement.

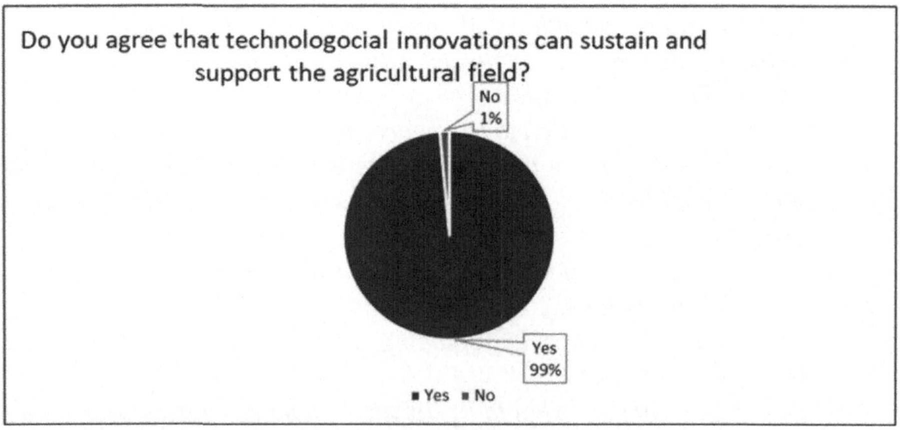

FIGURE 5.31 Respondents' opinions on technological innovations can sustain and support the agricultural field.

In Figure 5.32, the researcher also asked the respondents about what kinds of technological innovations can sustain and support the agricultural field. Approximately 115 respondents, accounting for 24%, agreed to using the sun to generate freshwater. Another significant modern-day agricultural trend is to increase water productivity in agriculture and food production. This represents a critical metric in the face of global population growth and climate change, whether for traditional agricultural irrigation, arid areas, or urban farming. Followed by 99 respondents, 21% agreed to innovative agriculture moving underground. It utilizes hydroponic systems and LED lighting to cultivate a variety of herbs and vegetables, including celery, rocket, parsley, radish, and mustard leaf. About 95 respondents, 20% agreed to preserving data in soil. The availability and suitability of land are perhaps the most significant factors determining supply efficiency in traditional farming practices. However, any assessment of future potential should be based on current data, which can then be used to inform the narrative of a region. In turn, this story is written in the earth, the soil. Around 82 respondents, or 18%, agreed to the consumption of greens made from rainbow waste. One common innovation in agriculture is the use of hydroponics, a growing technique that utilizes minerally enriched water. Aquaponics takes it a step further by combining fish and plant farming in one system. The fish, through aquaculture effluent, provide a significant portion of the plant nutrition. Fish waste metabolites are effectively removed through nitrification and then absorbed by the plants, with the properly treated water flowing back to the farm. About 79 respondents, or 17%, agreed that agriculture by aircraft can sustain and support agricultural fields. Aircraft designed specifically or adapted from a general utility airframe are used for agriculture, forestry, and insect vector control for human, animal, and plant diseases.

Furthermore, another technological innovation that could sustain and support the agriculture field in Brunei Darussalam is the Internet of Things (IoT). It utilizes sensors to collect data on soil, humidity, temperature, and other conditions that contribute to maintaining a healthy agricultural environment. Additionally, sensors are used to detect any anomalies in real-time. Moreover, IoT works to automate manual labor work in the field, such as irrigation. These readings trigger relevant actuators and authorities to take immediate actions. Big Data, which is equivalent to the collected data, is analyzed to discover hidden information and patterns that could help understand the current agricultural situation. Artificial intelligence (AI) also plays a significant role. AI is used to predict certain scenarios so that preliminary actions can be taken and preparations can be made in advance. It aids in understanding the growth of plants and the early detection of plants with diseases. Organic farming has been proven to increase yield without harming the environment. Therefore, farmers can increase yield in a more sustainable way. Automated farming and smart data systems are one way to ensure accurate investment and output. All of these technological approaches can greatly improve productivity. Technology has the potential to assist farmers in terms of accessing profitable markets, finance, and quality inputs at the right times of the season.

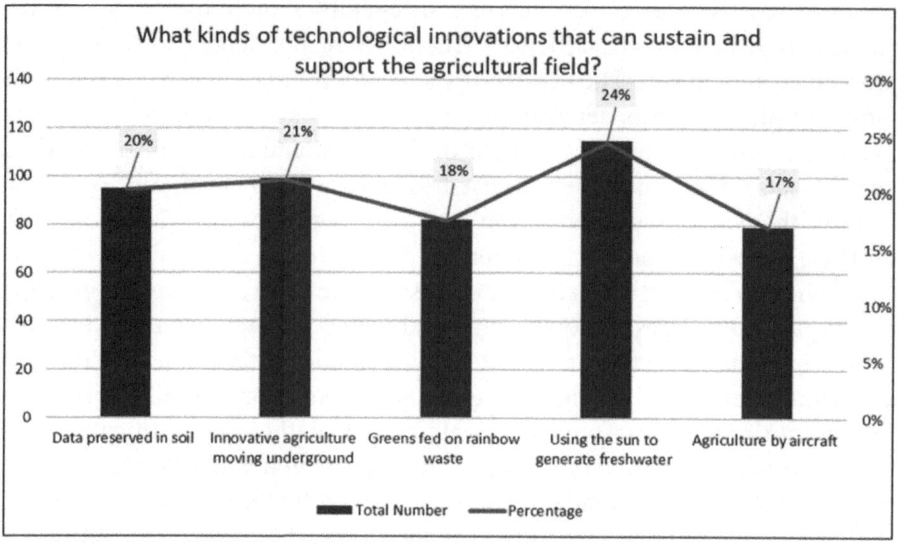

FIGURE 5.32 Technological innovations that can sustain and support the agricultural field.

5.6 CONCLUSIONS

Based on the study's results, it can be concluded that the findings of the studies are able to answer the important questions related to the goals of the research. The researcher aims to understand the extent of teenagers' exposure to the agriculture field in Brunei Darussalam. Through statistical analysis, respondents' opinions and theories about the agricultural field, a relationship between youth awareness and agriculture can be observed. It is important for youth to be more exposed to this field as they represent the future. Agriculture helps to diversify the economy and ensure survival, rather than solely relying on government jobs. They must be exposed to the agriculture sector at an early stage so that young technology can assist the older generation in enhancing agriculture with advanced technology. By exposing them to the industry, some may become interested in studying it further, leading to commercial purposes and increasing knowledge about agriculture, resulting in a wider range of agricultural products. This is important because not many people are currently exposed to agriculture, and it would be beneficial if they were introduced to this sector. Their involvement could help improve Brunei's production, thus enhancing the country's economy and reducing unemployment rates. Implementing technology in farming methods and encouraging fresh ideas in the industry would attract young people and their skills, contributing to the development of the rural sector.

5.7 CHALLENGES

Due to the outbreak of the COVID-19 pandemic, the researcher is encountering difficulties in terms of data collection and fieldwork. It is challenging for the researcher to collect data through face-to-face interviews, so they have resorted to conducting remote interviews via online conference platforms such as Zoom and Skype during the data collection process. In terms of time consumption, collecting the data for the report is time-consuming. Several procedures need to be taken to extract data from the private sector, such as preparing a project proposal or letter, obtaining approval, and initiating contact and meetings with the interviewees to discuss agricultural-related matters. Another difficulty is the limitation imposed by internet availability, which hinders the data collection process.

5.8 RECOMMENDATIONS

The study has shown that both government and non-government sectors need to establish and provide farming awareness workshops, seminars, and training programs to emphasize the value of agriculture. Additionally, the development of software to assist with farm work is crucial. It is important to educate individuals about the benefits of joining the agriculture sector, highlighting its lucrative and sustainable nature due to the consistent demand. To attract young people, initiatives such as creating supported ads on social media and implementing intensive marketing strategies, particularly from the agricultural department, must be undertaken. It is essential to promote the diverse range of opportunities available for young individuals, encouraging them to enhance their awareness and active participation in the field. When people experience its taste, its value will increase. The youth also need to change their mindset. The participation of the youth will contribute to the growth of Brunei's economy. Actions that the youth can take include demonstrating interest in this field. They need to be more committed, active, and interested in this field. There is a need for more publicity to help cultivate their interest in the field of agriculture.

KEYWORDS

- **agriculture sector**
- **big data**
- **Brunei's initiatives**
- **descriptive statistics analysis**
- **digital ecosystem**
- **security adoption**
- **statistical analysis**
- **technological innovations**
- **unemployment**

REFERENCES

1. Pehin, D. M. S., Pg Hj Idris, P., & Basir, K., (2020). *Exploring The Entrepreneurial Motivations and Barriers of Agripreneurs in Brunei Darussalam.* [online] Research Gate. Available at: https://www.researchgate.net/publication/338661819_Exploring_the_Entrepreneurial_Motivations_and_Barriers_of_Agripreneurs_in_Brunei_Darussalam (accessed on 16 January 2023).
2. Azlan, O., (2017). *Brunei to Post Positive Growth in 2018: ADB* – Borneo Bulletin Online. [online] Available at: https://borneobulletin.com.bn/brunei-to-see-growth-this-year-says-minister/ (accessed on 16 January 2023).
3. Koo J. S. (2013). Lowly educated youth a national concern. The Brunei Times, 30 October. Available at: http://www.bt.com.bn/news-national/2013/10/30/lowly-educated-youth-national concern (accessed on 5 April 2018).
4. *Joblessness Still Pains Brunei*, (2013). Available at http://www.asianewsnet.net/Joblessness-still-pains-Brunei-51468.html (accessed on 16 January 2023).
5. Bostyn, A. M., & Wight, D., (1987). Inside a community: Values associated with money and time. In: Fineman, S., (ed.), *Unemployment: Personal and Social Consequences* (pp. 149–151). London, England: Tavistock Publications Ltd.
6. Wong, A., (2019). *Inside The Home-Based Aquaponics Farm Shortlisted for Shell Global's Innovator Awards* | Biz Brunei. [online] Biz Brunei. Available at: https://www.

bizbrunei.com/2019/09/home-based-aquaponics-farm-shortlisted-of-shell-global-brunei-sr-aquafarm/ (accessed on 16 January 2023).

7. Mohamad, L., (2019). *Kemari Farm, Benchlab Bag Top Prizes*. Borneo Bulletin Online. [online] Borneo Bulletin Online. Available at: https://www.bizbrunei.com/2019/11/kemari-farm-wins-10000-from-scb-through-scot-programme/ (accessed on 16 January 2023).
8. Worldometer.info (2019). Obtained from: Worldometer – real time world statistics (worldometers.info).
9. Department of Economic, Planning, and Development. (2014). Obtained from: Home – Department of Economic Planning and Statistics (mofe.gov.bn).
10. World Bank. (2018). Obtained from World Bank Group - International Development, Poverty, & Sustainability.
11. Brunei's GDP for Agriculture. (2019). Obtained from: Brunei GDP (tradingeconomics.com).
12. Department of Agriculture and Agri-Food. (2018). Obtained: Home – Agriculture and Agrifood Department.

CHAPTER 6

Measurement, Modeling, and Sustainable Control of Environmental Contaminants Using Smart Computational Technologies: A Case Study

KIAN HARIRI ASLI

Department of Electronic Engineering, University of Rome "Tor Vergata," Rome, Italy

ABSTRACT

The present work deals with the control of environmental contaminants caused by toxic agents in asphalt. Natural asphalt produces desirable pavement using natural, non-toxic, readily available, and inexpensive materials. The identification of critical zones requiring pavement improvement was conducted through remote sensing (RS) and the internet of things (IoT). This study demonstrated that networked sensors, GIS, RS, and IoT enabled the prediction of the mechanical behavior of RAP in a short amount of time. Using GIS, networked sensors, RS, and IoT, the RAP was identified within a second. The present work compared the pavement data for the optimal bitumen percentage (OBP) with the dependent variable "Y" representing nomenclature and the independent variable "X" representing Stability The results demonstrated that the addition of rhamnolipid biosurfactant, a natural and non-toxic agent, effectively reduced fatigue susceptibility and enhanced control of environmental contaminants in asphalt.

Sustainability in Energy and Environment: Engineered Materials and Smart Computational Techniques.
Ann Rose Abraham, Heru Susanto, A. K. Haghi, & Kian Hariri Asli (Eds.)

6.1 INTRODUCTION

The Internet of Things (IoT) includes any device that can be turned on and off with a switch and can connect to the Internet. The IoT encompasses equipment that connects information through the Internet, enabling effective and seamless communication between them. One crucial aspect of this technology is that the history of the IoT is relatively short, spanning only a few years. However, the concept of machines communicating directly with each other dates back to the development of the telegraph in the 1830s and 1840s.

The first radio transmission, known as the "wireless telegraph," was produced on June 3, 1900, and provided the necessary components for the development of the IoT. This continued with the development of computers in the 1950s, which we will discuss in detail below.

In the first part of the history of the IoT, it should be noted that the development of the Internet itself is a very important part of the history of IoT technology. The Internet first started its work in 1962 and was officially introduced to the world in 1969. It was in the 1980s that commercial service providers began to support the public use of this technology and allowed it to evolve into the modern Internet of today. Satellites and fixed lines provided the initial communications in the context of the IoT in today's form.

In addition, GPS satellites became a reality in early 1993, providing a stable and highly functional system that contributed to the development of today's Internet. Furthermore, the placement of commercial satellites with private properties accelerated the growth of the Internet, creating a suitable platform for the IoT.

Today, computers and consequently the Internet are heavily reliant on humans for information. Almost all information on the Internet was first typed or stored by humans. This implies that humans are not particularly adept at gathering information from the physical world. Therefore, we require technologies that can gather information with the utmost accuracy and in the shortest amount of time possible.

Kevin Ashton was the person who first introduced the term IoT. He believed that RFID identification is a prerequisite for the IoT, and this issue has been repeated many times throughout the history of the IoT. He believed that if all devices were tagged, computers could manage, track, and control them. Today, to a large extent, this labeling operation can be done using technologies such as digital watermarks, barcodes, and QR codes [1–8].

In 2002, the first technology in the field of IoT was developed. This technology was a smart doorbell, which served as an excellent example of

IoT. This breakthrough occurred in 2011 and can be considered as the initial version of modern IoT products. Throughout the history of IoT, we have witnessed fascinating applications. One of the most significant demands in the world of IoT is the development of smart cities. In 2012, the Swiss Federal Energy Administration initiated an experimental program called smart city, aiming to utilize IoT technologies for city management. Traffic monitoring, weather monitoring, smart transportation, smart parking, and more were among the key features of this city.

With the passage of time in the history of IoT, we have witnessed the emergence of the Industrial IoT. In this phase, smart actuators and sensors were utilized for industrial applications, aiming to enhance efficiency and industrial capabilities.

6.1.1 THE IMPORTANCE OF USING THE INTERNET OF THINGS (IOT) IN TRANSPORTATION

In general, the transportation and logistics industry has always faced numerous challenges since ancient times and has been seeking effective solutions to overcome them. Some of the most significant challenges include:

- Having increased visibility in transportation processes;
- Operational performance measurement;
- Reduce downtime;
- Reduce operating costs;
- Optimizing road traffic.

Today, IoT technology can have a significant impact on overcoming these challenges and can reduce or even eliminate them in some cases. The utilization of IoT in transportation enables industries and organizations to personalize their products, operate more efficiently, and consequently reduce their overall costs.

6.1.2 SMART TRANSPORTATION THROUGH THE INTERNET OF THINGS (IOT)

In general, the IoT, due to its ability to connect different devices to each other, can be a suitable key for the effective transformation of logistics processes throughout the country.

6.1.3 ADVANTAGES OF USING THE INTERNET OF THINGS (IOT) IN TRANSPORTATION

Internet of Things (IoT)-based transportation can bring many benefits, including:

1. **Communication System:** An effective communication system between truck and car drivers, as well as managers, can greatly enhance their performance.
2. **Location Monitoring:** Today, modern technologies such as GPS are used to monitor the current location of vehicles. These equipment ensure that the intended cargo reaches its destination and is delivered at the specified time. This feature also allows logistics companies to estimate delivery times and adjust their vehicles accordingly.
3. **Supply Chain Monitoring:** The logistics industry is supported by real-time supply chain monitoring systems. These systems can easily perform all simplification operations in supply chain processes. To become familiar with the intelligent supply chain, we suggest reading Infotech's article in this field.
4. **Cyber Security:** Today, security threats are a major concern in every industry, including the transportation industry. The utilization of IoT in transportation can significantly contribute to the mitigation of these security threats.

6.1.4 APPLICATION AND BENEFITS OF INTERNET OF THINGS (IOT) IN INTELLIGENT TRANSPORTATION

6.1.4.1 IOT APPLICATIONS IN THE TRANSPORTATION INDUSTRY

Among the most important applications of the IoT in transportation, the following can be mentioned:

1. **Efficient Traffic Management:** Traffic management is the largest part of the transportation industry, which can be optimally achieved today through the utilization of IoT technologies.
2. **Issuing Tickets and Tolls Fully Automatically:** Traditional ticketing and toll collection systems are not only outdated, but they also do not contribute to the current flow of vehicles on the roads. This operation can be done in the best possible way by utilizing the technologies that have been designed and produced in the field of IoT, which today have wide applications in the transportation industry.

3. **Self-Driving Cars:** Undoubtedly, self-driving cars can be considered a real and very attractive example of the applications of the IoT in smart transportation. These cars are able to move easily in the environment without needing to have a driver.

By utilizing IoT equipment, it is possible to conduct vehicle tracking operations in a highly advanced manner and maintain full supervision over the transportation of various products [9–18].

6.1.4.2 SMART LOGISTICS

Smart logistics can be considered as a combination of traffic management structure and guidance for the optimal use of traffic systems. On the other hand, smart logistics in manufacturing industries is also defined by the concept of organizing, planning, controlling, and implementing the flow of goods in an effective way, with optimal use of data. To learn more about the concept of smart logistics, reading our article will be very helpful.

IoT has many applications in the construction and maintenance of roads and transportation. For example, we can mention the restoration of roads and optimal traffic management. The use of IoT equipment can help reduce costs in the transportation industry, optimizing the operations of moving different products as much as possible. For this reason, many investments have been made in the use of IoT in smart modeling and sustainable control of environmental contaminants using smart computational technologies for the maintenance of roads.

A Geographic Information System (GIS) is a computerized information system that typically produces, processes, analyzes, and manages geographic information. In other words, GIS is a computer system used to manage and analyze geographic information, with the ability to collect, store, analyze, and display such data. The ultimate goal of GIS is to support decision-making based on geographic data. Its basic function is to obtain information by combining different layers of data using various methods and perspectives. Additionally, manually digitized maps can also be used as GIS layers.

This work shows that GIS is used for grading and maintaining roads, managing transportation and urban traffic, and preparing basic road maps and measurements. It also involves modeling and sustainable control of environmental contaminants using smart computational technologies. The GIS database is georeferenced. The following list includes some of the most well-known applications of GIS in other industry fields [19–28]:

1. **Road Mapping and Maintenance:** The basic application of these standards and tools is in mapping, preparation of urban and suburban maps, and maintenance of roads.
2. **Earth Sciences:** One of the most important practical issues in GIS for earth sciences is performing overlaps to determine zoning, which is briefly addressed as one of these types of zoning.

The purpose of this zoning is to determine areas with landslide risk. To achieve this goal, one must first identify the factors that contribute to landslides. Among these factors, we can mention vegetation density, land type, topographic slope, slope of land layers, and rainfall intensity.

After identifying the factors, the required maps must be prepared. These maps can be created by overlapping the layers in the GIS software, resulting in a map that divides the studied area into smaller areas. In the database, the characteristics of each area, which are determined by various factors, are recorded. Each factor must be assigned a relevant weight in the database, and the final numbers can be obtained by adding and multiplying the factors. In the next step, it should be determined which number represents the high-risk areas that need to be marked as RAP. After classifying the map, this work investigates the areas with a field visit and modifies the final map to control environmental contaminants caused by asphalt.

This work determined the maximum amount of RAP, regardless of the type of aggregate it contained, that could be used in surface mixes.

The process includes the following items:

- Samples of RAP were collected from 10 different locations and analyzed to determine their binder content and aggregate gradation. The goal was to assess the overall variability of RAP.
- Using historical data and guidance from the Soil Mechanics laboratory.
- RAP samples were collected from various stockpiles to determine the aggregate gradation and binder content. The RAP stockpiles consisted of material milled off of various state and local contracts, which was then crushed and screened through a 25 mm [29–38].

6.2 METHODS

In this work, the following hypothesis can be expected from asphalt containing modified non-toxic bitumen:

- Does the use of modified bituminous surfactant rhamnolipid increase the resistance of asphalt samples to fatigue-induced cracks?

- Is it possible to identify the critical zone with the most needs for pavement using GIS, RS, and IoT?
- Why rapid data intercommunication technique be able to find critical zone with the most needs to the RAP up to one second?
- How can conceptual modeling predict the mechanical behavior resulting from RAP? In this study, the research methodology is presented in the form of a chart in Figure 6.1. The materials used include new communication techniques, laboratory equipment, and tests conducted on bitumen and asphalt samples.

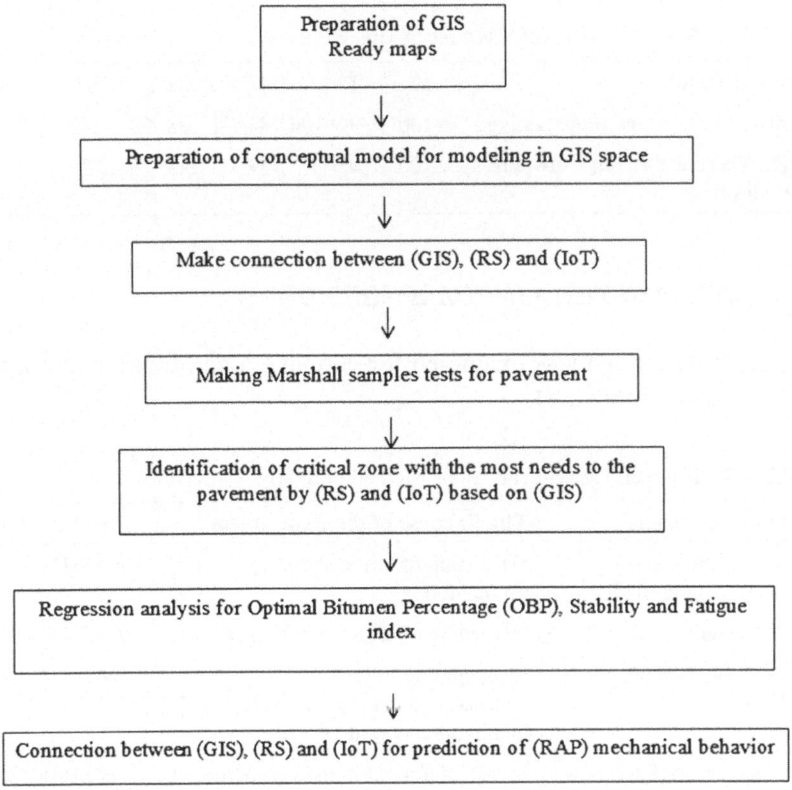

FIGURE 6.1 Research flow chart for controlling contaminants in asphalt.

6.2.1 TESTS OF STONE MATERIALS

In this work, the materials used in the construction of the asphalt mixture were subjected to granulation tests, specific gravity measurements, Los Angeles wear tests, and fracture percentage evaluations. These tests were conducted on aggregates prepared from the mine, ensuring that they meet the necessary specifications.

6.2.2 SEEDING

The granulation of stone materials used in this work involves preparing a sample of asphalt mixtures. The continuous granulation number is 4 out of 234 (Table 6.1).

TABLE 6.1 Sample of Asphalt Mixtures in This Work

Sieve Size (mm)	19	12	4.75	2.36	0.3	0.0075
Permitted Pass Percentage Range	100	90–100	44–74	28–58	5–21	2–10
Weight Percent Passing Through Each Sieve	100	95	59	43	13	6

6.2.3 BITUMEN TESTS IN SHARP METHOD

The selected bitumen was examined for permanent deformation and fatigue crack increase (Table 6.2).

TABLE 6.2 Bitumen Experiments in the Method of Sharp

Type of Experiment	The Purpose of the Experiment	Test Standard
Test of thin bitumen by rotational method (RTFO).	Hardening of bitumen during asphalt production.	AASHTO-T240
Aging acceleration test (PAV).	Hardening of bitumen during service.	AASHTO-PP1
Dynamic shear rheometer (DSR) test.	Investigation of groove properties and cracks due to fatigue at high and medium temperatures.	AASHTO-TP5
Rotational chord (RV) test.	Investigation of bitumen properties at high temperature–bitumen efficiency.	AASHTO-D4402
Bending beam rheometer (BBR) test.	Investigation of bitumen properties in low temperature cracks.	AASHTO-TP1
Direct traction test (DT).	Investigation of bitumen properties in low temperature cracks.	AASHTO-TP3
Bitumen flash point test.	Determine the flash point of bitumen.	AASHTO-T48

6.2.4 RESEARCH TOOLS

The research tools included networked sensors, geospatial information system (GIS), remote sensing (RS), and IoT for contaminant control of asphalt.

6.2.5 NETWORKED SENSORS

Networked sensors and IoT based on GIS are a collection consisting of components of information and data, hardware, software, power, and expert users. They work together to collect, enter, store, evaluate, retrieve, manage, analyze, and display descriptive and spatial data. GIS is used in various sciences and technologies, including mapping, road networks, transportation, environment (control of environmental contaminants due to toxic agents in asphalt), crisis management and unexpected events, urban and rural development, different agricultural sectors, different water engineering sectors, hydrology, hydrogeology, hydraulics, and planning of water resources, forests, insurance industry, social and economic sciences, and project documentation construction [3, 4, 10, 39–50].

6.2.6 GEOSPATIAL INFORMATION SYSTEM (GIS) AND REMOTE SENSING (RS)

To display spatial data in GIS, various structures such as vector, raster, and cellular are utilized. Vector and raster structures are the primary methods for representing spatial information. In the vector structure, phenomena in the natural world are represented using points (e.g., well locations), lines (e.g., roads, rivers), and polygons (e.g., forests, pastures, fields, bodies of water). On the other hand, the grid or cellular structure represents phenomena and complexities as a grid of cells, with each cell corresponding to a specific row and column position. This structure is commonly used for displaying images, satellite data, digital elevation models (DEM), and scanned maps [5–11, 51–53].

The TIFF format is used in most GIS and RS software. Image processing software can produce and support JPG. Raster data with this format has the ability to maintain the coordinate system. They don't have their own, and in fact, they are not the ground of reference. Usually, digital photos or scanned paper maps are saved in such formats. Compared to other formats, JPG has a lower file size.

Hardware and software are important components of GIS. In recent decades, along with the development of computer hardware, software has also been developed.

The development and expansion of software are always one step ahead of hardware developments. Since the formation of GIS until now, various software has been prepared and presented. Some of the most famous ones are ILwis, ARC-GIS, Idrisi, and Arc/Info. The main structure of some of these software, like ARC View and GIS, is vector ARC, while others like ILwis and Idrisi have a dominant grid structure. Without a doubt, ARC-GIS, marketed by Esri, is the most complete and comprehensive software available. ARC-GIS is included in a collection of software packages: Arc-Reader, Arc-Scan, Arc-Object, Arc-Globe, Arc-Catalog, Arc-Map, and Arc-Scene. All of these software work together and form an application that can be linked to ARC-GIS. So far, the versions of ARC-GIS include ARC-GIS 10, ARC-GIS 9.3, ARC-GIS 9.2, and ARC-GIS 9.1 [12, 13, 15–17, 19, 27, 54–56].

6.2.7 INTERNET OF THINGS (IOT)

At the beginning of the 21st century, the possibilities that the IoT offers to human society are endless. The data collected by all IoT devices and sensors has increased and provides big data. IoT is generating a tsunami of big data, which is intensifying with the rapid expansion of devices and sensors. The amount of information generated by IoT is increasing astronomically. This data captures valuable insights into what is working well. Some of the capabilities of IoT data are [20, 21, 23–26, 28, 57–59]:

- Control of environmental contaminants due to the toxic agents in asphalt.
- Research for locations that require natural asphalt pavement using natural, non-toxic, readily available, and inexpensive materials.
- Investigation of the effects of rhamnolipid biosurfactant on improving the technical properties of asphalt pavement for the control of non-toxic agents.
- Urban management to predict accidents and crimes.
- Understanding the patients' conditions, IoT provides vital information to doctors in the shortest possible time.
- This helps improve productivity in industries by performing scheduled PM maintenance for equipment.

- The devices and sensors installed in the smart home assist in the efficient management of residential and industrial spaces.
- Provides communication with the driver's camera.

Currently, humans can easily use and understand all the data from traditional and modern methods. The big problem is finding a way to analyze the performance data and the information generated by these devices. Understanding data in terabytes is a significant challenge that only IT scientists can comprehend. Therefore, the full advantages of IoT are as follows:

- High speed analysis of big data;
- Accurate analysis of big data.

6.2.8 ARTIFICIAL INTELLIGENCE (AI), INTERNET OF THINGS (IOT), AND GEOSPATIAL INFORMATION SYSTEM (GIS)

The rapid circulation of location-based information in the Internet space of IoT objects, under the management of artificial intelligence (AI), leads to the optimization of environmental management through remote reading technology. The execution of these operations will provide an intelligent management model.

Geographic information system (GIS) – it is a system consisting of data, hardware, software, algorithmic methods, human power, and a network that has the ability to input, manage, analyze, and display "spatial information." The components of the phrase are:

- Complications: Spatial feature is a spatial phenomenon;
- Information: Information shows processing data;
- Human power: Personnel with dynamic thinking are the main key to the power of GIS;
- System: System that establishes the connection between software, hardware, and data.

The global Internet network is a GIS-based Web type of GIS that enables the distribution, sharing, and exchange of data at any time, any place, and for any person and problem through the global Web network. Therefore, LINE-ON data analysis through the technology of signal reception in a fraction of a second through sensors equipped with remote reading modems via the global Internet network can lead to line control and management of water and energy consumption. Control science leads to the control of environmental contaminants due to the toxic agents in asphalt and optimization of energy consumption through the integration of sensor systems connected to the Internet.

- Integrating the Internet with IoT-based building water and energy management systems in "smart building units."
- Integration of smart equipment is aimed at enabling fast intercommunication of location-based information within the Internet space of IoT objects, all under the management of AI.

6.2.9 REGRESSION ANALYSIS

Regression builds a function that shows the interaction between parameters.

6.3 RESULTS AND DISCUSSION

6.3.1 RESULTS OF MARSHALL TESTS FOR ASPHALT SAMPLES

Data analysis and management with the help of GIS is easily and efficiently conducted. In this work, GIS is utilized to create databases for controlling environmental contaminants caused by toxic agents in asphalt. The data and information in GIS are used both descriptively and spatially.

Descriptive information shows the nature of complications and phenomena, providing georeferenced data on roads (Tables 6.3–6.8 and Figures 6.2–6.9).

TABLE 6.3 Graphs of the Results of the Marshall Test for the Control Sample: Unit Weight

Unit Weight (Kg/m3)	Asphalt Content (%)
2,220	4
2,240	4.5
2,260	5
2,280	5.5
2,300	6
2,320	6.5
2,340	7
2,360	7.5

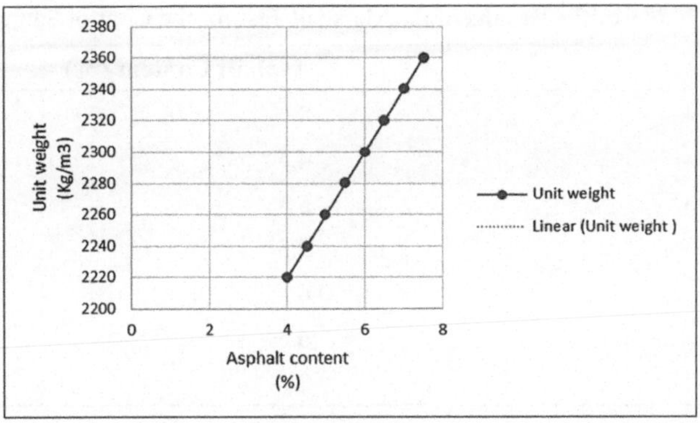

FIGURE 6.2 Graphs of the results of the Marshall test for the control sample; unit weight.

TABLE 6.4 Graphs of the Results of the Marshall Test for Control Sample; Air Voids

Air Voids	Asphalt Content (%)
0	0
1	7
2	6.8
3	6.5
4	6.2
5	5.8
6	5.2
7	5.3
8	5
9	4.7
10	4.5
11	4.5

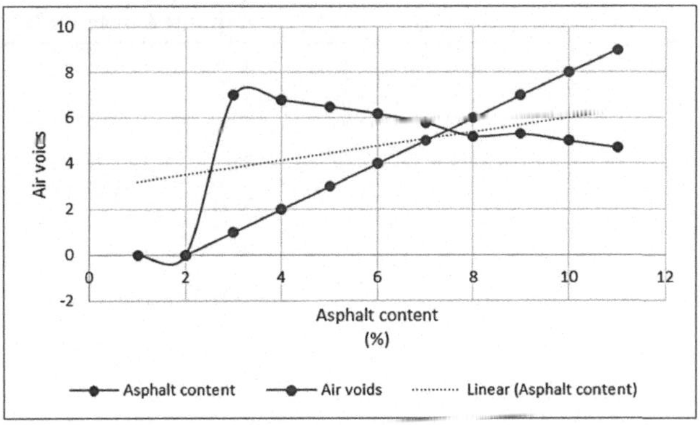

FIGURE 6.3 Graphs of the results of the Marshall test for the control sample: air voids.

TABLE 6.5 Graphs of the Results of the Marshall Test for the Control Sample: Stability

Stability (%)	Asphalt Content (%)
700	4
800	4.5
900	5
1,000	5.5
1,100	6
1,200	6.5
1,300	7

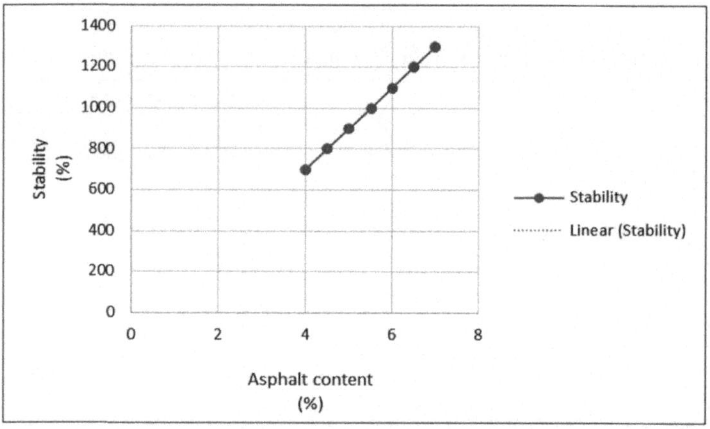

FIGURE 6.4 Graphs of the results of the Marshall test for the control sample; stability.

TABLE 6.6 Graphs of the Results of the Marshall Test for the Control Sample: VMA

VMA (%)	Asphalt Content (%)
14	4
15	4.5
16	5
17	5.5
18	6
19	6.5
20	7
21	7.5
22	8

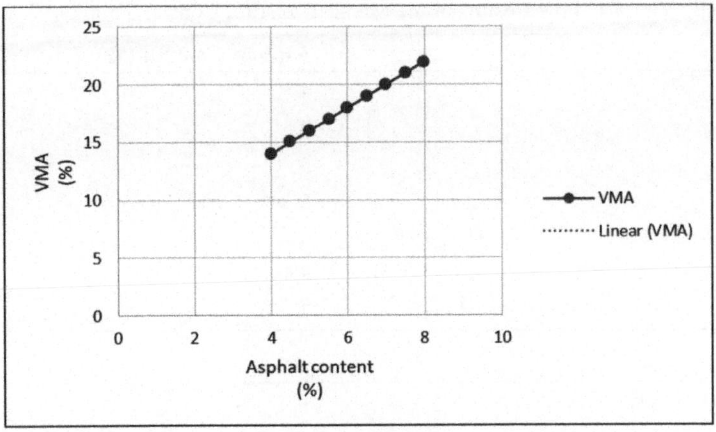

FIGURE 6.5 Graphs of the results of the Marshall test for the control sample: VMA.

TABLE 6.7 Graphs of the Results of the Marshall Test for Control Sample: Flow

Flow	Asphalt Content (%)
9	4
10	4.5
11	5
12	5.5
13	6
14	6.5
15	7

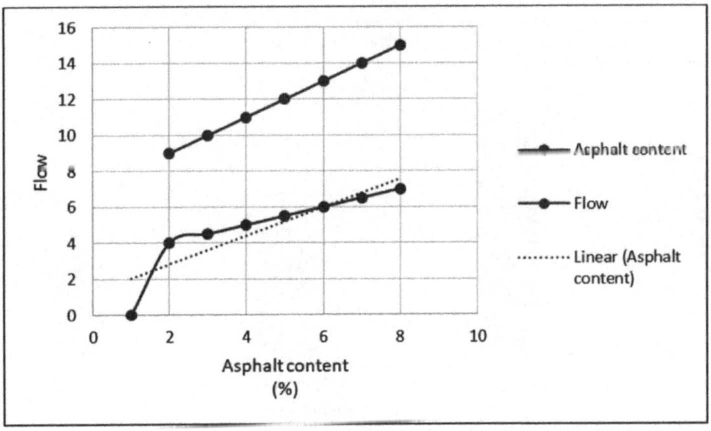

FIGURE 6.6 Graphs of the results of the Marshall test for control sample; flow.

TABLE 6.8 Graphs of the Results of the Marshall Test for Control Sample: VFA

VFA (%)	Asphalt Content (%)
40	4
50	4.5
60	5
70	5.5
80	6
90	6.5
100	7

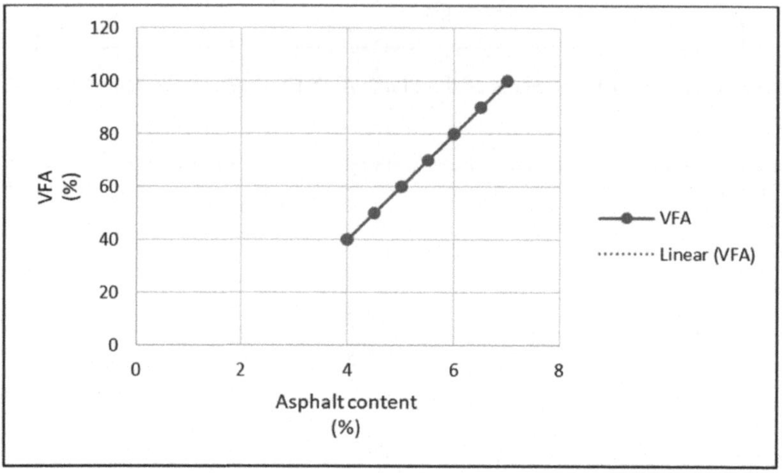

FIGURE 6.7 Graphs of the results of the Marshall test for control sample; VFA.

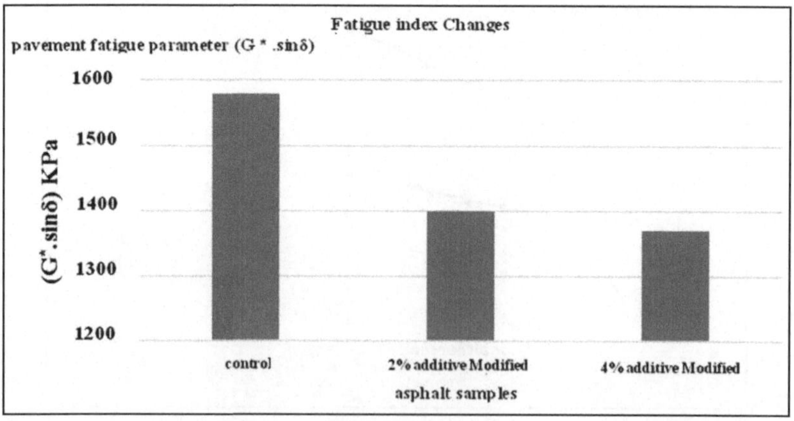

FIGURE 6.8 Fatigue index against different values of biosurfactant rhamnolipid.

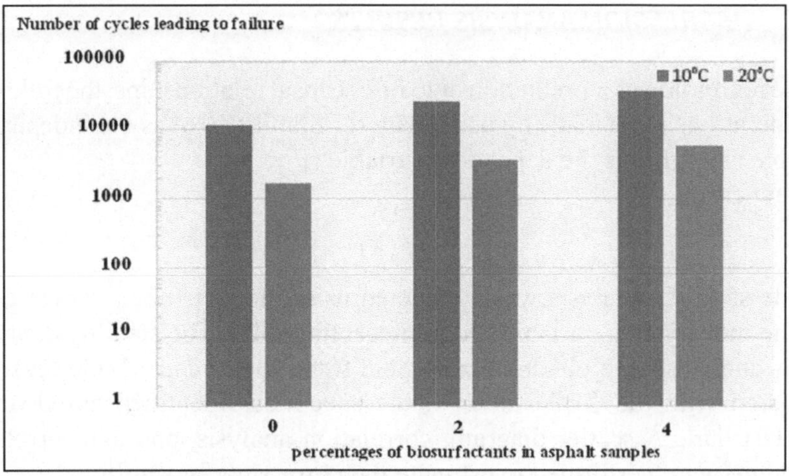

FIGURE 6.9 Fatigue test results for specimen.

6.3.2 *RESEARCH FORMULATION*

In this work, the following considerations were made using data for Stability (%) and fatigue index as independent variables with the nomenclature "X" against the optimal bitumen percentage (OBP) as the dependent variable with the nomenclature "Y" (Eqn. (1)):

$$Y = f(X_1, X_2) \tag{1}$$

The curve estimation procedure was performed using the "Y" and "X" data that were obtained from the field tests. The scatter diagram and correlation analysis were conducted to estimate the regression statistics. The calculations for the three cases were based on the experiments and laboratory tests, as shown in Equations (2) and (3).

First case:

$$O = f(S) \tag{2}$$

where; S is the stability of pilot as a field data; and O is the OBP.

Second case:

$$O = f(S, F.I.) \tag{3}$$

Where S represents the stability of the pilot as a field data, $F.I.$ represents the fatigue index, and O represents the OBP.

The regression and model summary indicate that the most significant effects are concentrated on Stability.

6.3.3 REGRESSION ANALYSIS RESULTS

To use regressions for prediction or to infer causal relationships, the following considerations were made by using data for Stability (%) as the independent variable and OBP as the dependent variable (Eqn. (4)):

First case:

$$Y = f(X) \tag{4}$$

The stability analysis was conducted using the dependent variable "Y" and the independent variable "X" representing OBP. The stability data used in this analysis were obtained from field tests. These data (Table 6.9) were compared with the OBP values of the specimens identified based on the field test data. A scatter diagram, correlation analysis, and autocorrelation procedure were performed to estimate the regression for stability.

TABLE 6.9 Graphs of the Results of the Marshall Test for Control Sample: Stability

Stability (%)	Optimal Bitumen Percentage (OBP)
700	4
800	4.5
900	5
1,000	5.5
1,100	6
1,200	6.5
1,300	7

In this work stability was based on the following case (Eqn. (5)):

- **In First Assumption for OBP Against Stability:** The curve estimation procedure allows for quick estimation of regression statistics and the production of related plots for different models.

The nomenclatures for the two cases are as follows:

- Dependent variable with nomenclature "*o*": OBP; and
- Independent variable with nomenclature "*s*": Stability.

Dependent variable with nomenclature "*O*": OBP, mm; independent variable with nomenclature "*S*."

$$O = f(S) \tag{5}$$

where; S is the stability of pilot as a field data.

The most important factors are as follows: concentration on the stability of the pilot. The curve fit (Figures 6.10–6.25) for dependent and independent variables was analyzed (Tables 6.10–6.77) [3–13, 15–17, 19–21, 23–64].

TABLE 6.10 NPar Tests: Notes

Output Created		**30-DEC-2022 11:29:55**
Syntax		NPAR TEST /CHISQUARE=VAR00001VAR00002 /EXPECTED=EQUAL /MISSING ANALYSIS.
Resources	Elapsed time	0:00:00.00
	Number of cases allowed[(a)]	1,04,857

[(a)] *Based on availability of workspace memory.*

TABLE 6.11 Chi-Square Test: Frequencies

	Optimal Bitumen Percentage (OBP)		
	Observed N	**Expected N**	**Residual**
4.00	1	1.0	0.0
4.50	1	1.0	0.0
5.00	1	1.0	0.0
5.50	1	1.0	0.0
6.00	1	1.0	0.0
6.50	1	1.0	0.0
7.00	1	1.0	0.0
Total	7		

TABLE 6.12 Stability (%)

	Observed N	**Expected N**	**Residual**
700.00	1	1.0	0.0
800.00	1	1.0	0.0
900.00	1	1.0	0.0
1000.00	1	1.0	0.0
1100.00	1	1.0	0.0
1200.00	1	1.0	0.0
1300.00	1	1.0	0.0
Total	7		

TABLE 6.13 Test Statistics

Parameters	Optimal Bitumen Percentage (OBP)	Stability (%)
Chi-square	0.000	0.000
df	6	6
Asymp. Sig.	1.000	1.000

TABLE 6.14 ACF Notes

Syntax		ACF VARIABLES = VAR00001 VAR00002/ NOLOG/MXAUTO 16/ SERROR=IND/PACF
Resources	Elapsed time	0:00:01.20
Use	From	First observation
	To	Last observation
Time series settings (TSET)	Amount of output	PRINT = DEFAULT
	Saving new variables	NEWVAR = NONE
	Maximum number of lags in autocorrelation or partial autocorrelation plots.	MXAUTO = 16
	Maximum number of lags per cross-correlation plots.	MXCROSS = 7
	Maximum number of new variables generated per procedure.	MXNEWVAR = 60
	Maximum number of new cases per procedure.	MXPREDICT = 1000
	Treatment of user-missing values.	MISSING = EXCLUDE
	Confidence interval percentage value.	CIN = 95
	Tolerance for entering variables in regression equations.	TOLER =.0001
	Maximum iterative parameter change.	CNVERGE =.001
	Method of calculating std. errors for autocorrelations.	ACFSE = IND
	Length of seasonal period.	Unspecified
	Variable whose values label observations in plots.	Unspecified
	Equations include.	Constant

TABLE 6.15 Model Description

Model Name		**MOD_2**
Series name	1	Optimal bitumen percentage (OBP)
	2	Stability (%)
Transformation		None
Non-seasonal differencing		0
Seasonal differencing		0
Length of seasonal period		No periodicity
Maximum number of lags		16
Process assumed for calculating the standard errors of the autocorrelations		Independence (white noise)[a]
Display and plot		All lags

Applying the model specifications from MOD_2.

[a]*Not applicable for calculating the standard errors of the partial autocorrelations.*

TABLE 6.16 Case Processing Summary

		Optimal Bitumen Percentage (OBP)	**Stability (%)**
Series length		7	7
Number of missing values	User-missing	0	0
	System-missing	0	0
Number of valid values		7	7
Number of computable first lags		6	6

TABLE 6.17 Optimal Bitumen Percentage (OBP) Autocorrelations

Series: Optimal Bitumen Percentage (OBP)					
Lag	**Autocorrelation**	**Std. Error**[a]	**Box-Ljung Statistic**		
			Value	**df**	**Sig.**[b]
1	0.571	0.309	3.429	1	0.064
2	0.179	0.282	3.830	2	0.147
3	–0.143	0.252	4.152	3	0.246
4	–0.357	0.218	6.830	4	0.145
5	–0.429	0.178	12.616	5	0.027

[a]*The underlying process assumed is independence (white noise).*

[b]*Based on the asymptotic chi-square approximation.*

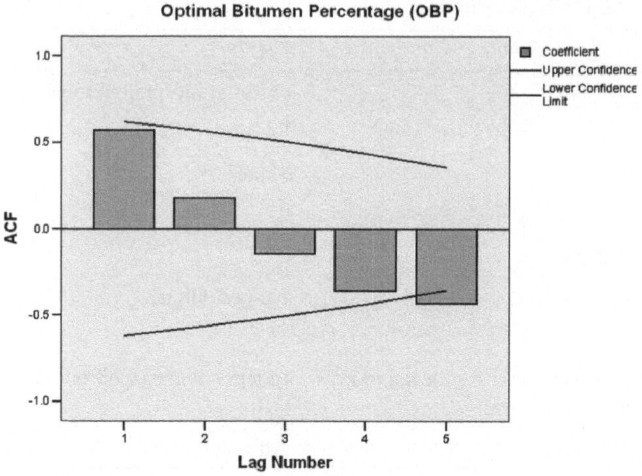

FIGURE 6.10 Autocorrelations for optimal bitumen percentage (OBP).

TABLE 6.18 Partial Autocorrelations

Series: Optimal Bitumen Percentage (OBP)		
Lag	**Partial Autocorrelation**	**Std. Error**
1	0.571	0.378
2	–0.220	0.378
3	–0.221	0.378
4	–0.207	0.378
5	–0.155	0.378

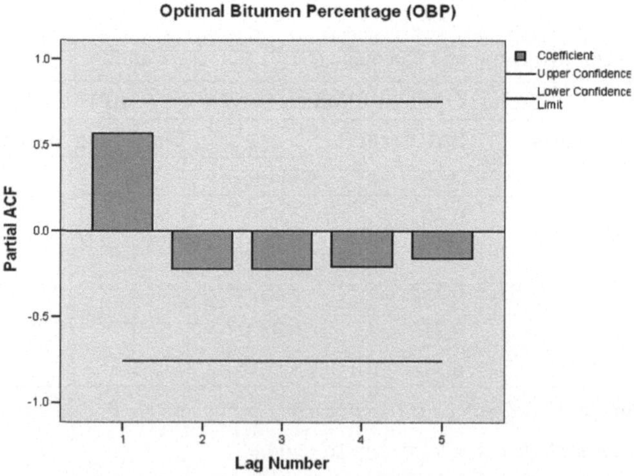

FIGURE 6.11 Autocorrelations for optimal bitumen percentage (OBP).

TABLE 6.19 Stability (%) Autocorrelations

Series: Stability (%)					
Lag	**Autocorrelation**	**Std. Error**[a]	**Box-Ljung Statistic**		
			Value	**df**	**Sig.**[b]
1	0.571	0.309	3.429	1	0.064
2	0.179	0.282	3.830	2	0.147
3	–0.143	0.252	4.152	3	0.246
4	–0.357	0.218	6.830	4	0.145
5	–0.429	0.178	12.616	5	0.027

[a] *The underlying process assumed is independence (white noise).*

[b] *Based on the asymptotic chi-square approximation.*

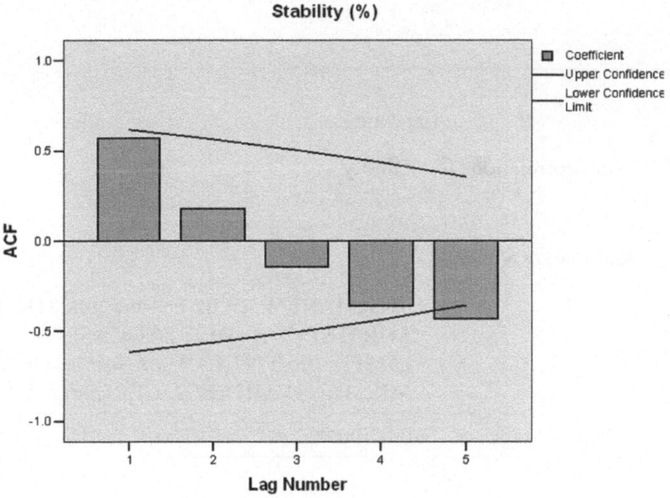

FIGURE 6.12 Autocorrelations for optimal bitumen percentage (OBP).

TABLE 6.20 Partial Autocorrelations

Series: Stability (%)		
Lag	**Partial Autocorrelation**	**Std. Error**
1	0.571	0.378
2	–0.220	0.378
3	–0.221	0.378
4	–0.207	0.378
5	–0.155	0.378

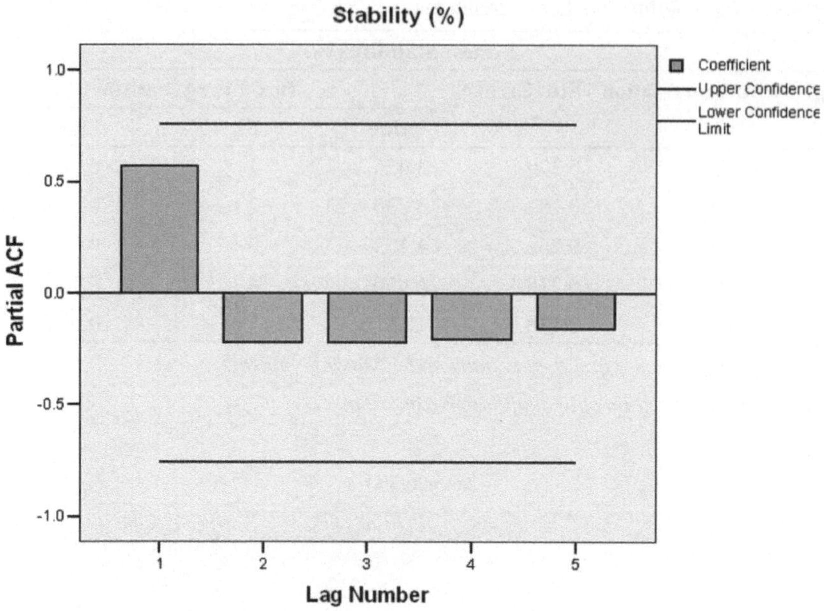

FIGURE 6.13 Autocorrelations for stability.

TABLE 6.21 Summarize Notes

Syntax		**SUMMARIZE/TABLES=VAR00001VAR00002/ FORMAT=VALIDLISTNOCASENUM TOTAL LIMIT=100/TITLE='Case Summaries'/ MISSING=VARIABLE/CELLS=COUNT.**
Resources	Elapsed time	0:00:00.00

TABLE 6.22 Case Processing Summary[a]

	Cases					
	Included		**Excluded**		**Total**	
	N	**Percent**	**N**	**Percent**	**N**	**Percent**
Optimal Bitumen Percentage (OBP)	7	100.0%	0	0.0%	7	100.0%
Stability (%)	7	100.0%	0	0.0%	7	100.0%

[a]*Limited to first 100 cases.*

TABLE 6.23 Case Summaries[a]

		Optimal Bitumen Percentage (OBP)	Stability (%)
1		4.00	700.00
2		4.50	800.00
3		5.00	900.00
4		5.50	1000.00
5		6.00	1100.00
6		6.50	1200.00
7		7.00	1300.00
Total	N	**7**	**7**

[a]*Limited to first 100 cases.*

TABLE 6.24 Logistic Model Summary

R	R Square	Adjusted R Square	Std. Error of the Estimate
0.997	0.994	0.992	0.018

Note: The independent variable is stability (%).

TABLE 6.25 ANOVA

	Sum of Squares	df	Mean Square	F	Sig.
Regression	0.241	1	0.241	766.057	0.000
Residual	0.002	5	0.000		
Total	**0.242**	**6**			

Note: The independent variable is stability (%).

TABLE 6.26 Coefficients

	Unstandardized Coefficients		Standardized Coefficients	t	Sig.
	B	Std. Error	Beta		
Stability (%)	0.999	0.000	0.369	29845.605	0.000
(Constant)	0.468	0.0160	–	29.266	0.000

Note: The dependent variable is ln(1/optimal bitumen percentage (OBP)).

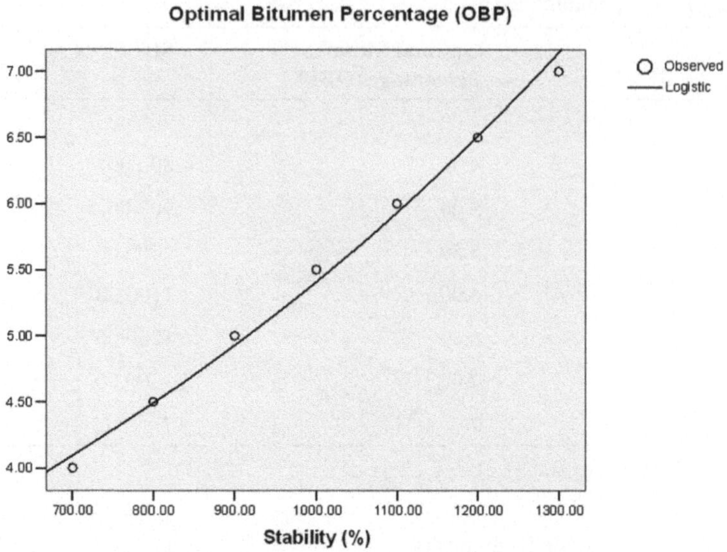

FIGURE 6.14 Scatter diagram for stability and OBP; logistic model.

TABLE 6.27 NPar Tests Notes

Output Created		**30-DEC-2022 13:55:42**
Comments		
Input	Filter	<none>
	Weight	<none>
	Split file	<none>
	N of rows in working data file	7
Missing value handling	Definition of missing	User-defined missing values are treated as missing.
	Cases used	Statistics for each test are based on all cases with valid data for the variable(s) used in that test.
Syntax		NPAR TEST / CHISQUARE=VAR00001VAR00002 /EXPECTED=EQUAL /MISSING ANALYSIS.
Resources	Elapsed time	0:00:00.01
	Number of cases allowed[a]	104857

[a]*Based on availability of workspace memory.*

TABLE 6.28 Chi-Square Test Frequencies

	Optimal Bitumen Percentage (OBP)		
	Observed N	**Expected N**	**Residual**
4.00	1	1.0	0.0
4.50	1	1.0	0.0
5.00	1	1.0	0.0
5.50	1	1.0	0.0
6.00	1	1.0	0.0
6.50	1	1.0	0.0
7.00	1	1.0	0.0
Total	7		

TABLE 6.29 Stability (%)

	Observed N	Expected N	Residual
700.00	1	1.0	0.0
800.00	1	1.0	0.0
900.00	1	1.0	0.0
1000.00	1	1.0	0.0
1100.00	1	1.0	0.0
1200.00	1	1.0	0.0
1300.00	1	1.0	0.0
Total	7		

TABLE 6.30 Test Statistics

Parameters	Optimal Bitumen Percentage (OBP)	Stability (%)
Chi-square[(a)]	0.000	0.000
df	6	6
Asymp. Sig.	1.000	1.000

[(a)] 7 cells (100.0%) have expected frequencies less than 5. The minimum expected cell frequency is 1.0.

TABLE 6.31 Graph Notes

Output Created		30-DEC-2022 11:20:30
	Comments	
Input	Filter	<none>
	Weight	<none>
	Split file	<none>
	N of rows in working data file	7
	Syntax	GRAPH /SCATTERPLOT(BIVAR) =VAR00002 WITH VAR00001/MISSING=LISTWISE.
Resources	Elapsed time	0:00:00.98

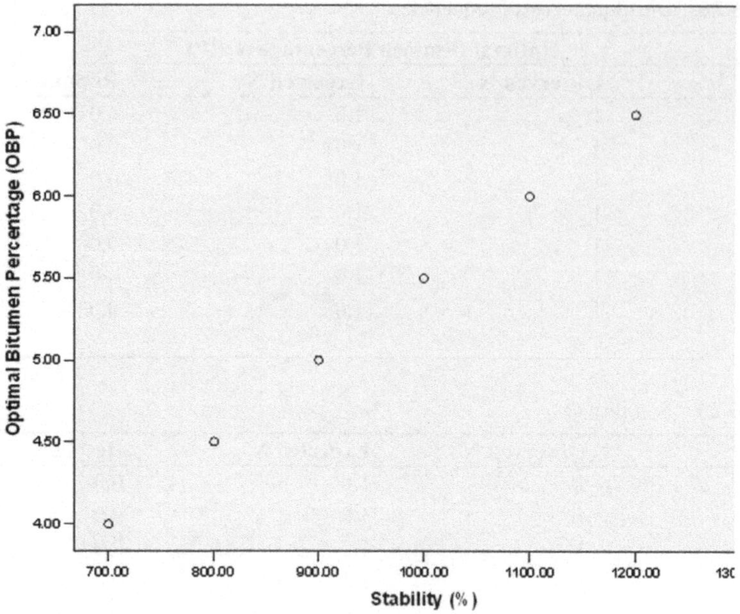

FIGURE 6.15 Scatter diagram for stability and OBP.

TABLE 6.32 NPar Tests Notes

Output Created		**30-DEC-2022 11:29:55**
	Comments	
Input	Filter	<none>
	Weight	<none>
	Split file	<none>
	N of rows in working data file	7
Missing value handling	Definition of missing	User-defined missing values are treated as missing.
	Cases used	Statistics for each test are based on all cases with valid data for the variable(s) used in that test.
	Syntax	NPAR TEST/CHISQUARE= VAR00001VAR00002/EXPECTED= EQUAL/MISSING ANALYSIS.
Resources	Elapsed time	0:00:00.00
	Number of cases allowed[a]	1,04,857

[a]*Based on availability of workspace memory.*

TABLE 6.33 Chi-Square Test Frequencies

Optimal Bitumen Percentage (OBP)			
	Observed N	**Expected N**	**Residual**
4.00	1	1.0	0.0
4.50	1	1.0	0.0
5.00	1	1.0	0.0
5.50	1	1.0	0.0
6.00	1	1.0	0.0
6.50	1	1.0	0.0
7.00	1	1.0	0.0
Total	7		

TABLE 6.34 Stability (%)

	Observed N	**Expected N**	**Residual**
700.00	1	1.0	0.0
800.00	1	1.0	0.0
900.00	1	1.0	0.0
1000.00	1	1.0	0.0
1100.00	1	1.0	0.0
1200.00	1	1.0	0.0
1300.00	1	1.0	0.0
Total	7		

TABLE 6.35 Test Statistics

Parameters	**Optimal Bitumen Percentage (OBP)**	**Stability (%)**
Chi-square[a]	0.000	0.000
Df	6	6
Asymp. Sig.	1.000	1.000

[a] 7 cells (100.0%) have expected frequencies less than 5. The minimum expected cell frequency is 1.0.

TABLE 6.36 NPar Tests Notes

Output Created		**30-DEC-2022 13:55:42**
	Comments	
Input	Filter	<none>
	Weight	<none>
	Split file	<none>
	N of rows in working data file	7
Missing value handling	Definition of missing	User-defined missing values are treated as missing.
	Cases used	Statistics for each test are based on all cases with valid data for the variable(s) used in that test.
	Syntax	NPAR TEST/CHISQUARE=VAR00001 VAR00002/EXPECTED=EQUAL/ MISSING ANALYSIS.
Resources	Elapsed time	0:00:00.01
	Number of cases allowed[a]	1,04,857

[a]*Based on availability of workspace memory.*

TABLE 6.37 Chi-Square Test Frequencies

	Optimal Bitumen Percentage (OBP)		
	Observed N	**Expected N**	**Residual**
4.00	1	1.0	0.0
4.50	1	1.0	0.0
5.00	1	1.0	0.0
5.50	1	1.0	0.0
6.00	1	1.0	0.0
6.50	1	1.0	0.0
7.00	1	1.0	0.0
Total	**7**		

TABLE 6.38 Stability (%)

	Observed N	**Expected N**	**Residual**
700.00	1	1.0	0.0
800.00	1	1.0	0.0
900.00	1	1.0	0.0
1000.00	1	1.0	0.0
1100.00	1	1.0	0.0
1200.00	1	1.0	0.0
1300.00	1	1.0	0.0
Total	**7**		

TABLE 6.39 Test Statistics

Parameters	Optimal Bitumen Percentage (OBP)	Stability (%)
Chi-square[(a)]	0.000	0.000
df	6	6
Asymp. Sig.	1.000	1.000

[(a)] 7 cells (100.0%) have expected frequencies less than 5. The minimum expected cell frequency is 1.0.

TABLE 6.40 Curve Fit Notes

Output Created		**07-JAN-2023 11:46:27**
	Comments	
Input	Data	C:\Users\taban\Desktop\BOOK-HOSSEIN-TOTAL\data-1.sav
	Filter	<none>
	Weight	<none>
	Split file	<none>
	N of rows in working data file	7
Missing value handling	Definition of missing	User-defined missing values are treated as missing.
	Cases used	Cases with a missing value in any variable are not used in the analysis.
	Syntax	CURVEFIT/VARIABLES=VAR00001 WITH VAR00002/CONSTANT/ MODEL= QUADRATIC/PLOT FIT.
Resources	Elapsed time	0:00:00.65
Use	From	First observation
	To	Last observation
Predict	From	First observation following the use period
	To	Last observation
Time series settings (TSET)	Amount of output	PRINT = DEFAULT
	Saving new variables	NEWVAR = NONE
	Maximum number of lags in autocorrelation or partial autocorrelation plots.	MXAUTO = 16
	Maximum number of lags per cross-correlation plots.	MXCROSS = 7
	Maximum number of new variables generated per procedure.	MXNEWVAR = 60
	Maximum number of new cases per procedure.	MXPREDICT = 1000
	Treatment of user-missing values.	MISSING = EXCLUDE

TABLE 6.40 *(Continued)*

Output Created		07-JAN-2023 11:46:27
	Confidence interval percentage value.	CIN = 95
	Tolerance for entering variables in regression equations.	TOLER =.0001
	Maximum iterative parameter change.	CNVERGE =.001
	Method of calculating std. errors for autocorrelations.	ACFSE = IND
	Length of seasonal period.	Unspecified
	Variable whose values label observations in plots.	Unspecified
	Equations include.	CONSTANT

TABLE 6.41 Model Description

Model Name		MOD_1
Dependent variable	1	Optimal bitumen percentage (OBP)
Equation	1	Quadratic
Independent variable		Stability (%)

TABLE 6.42 Case Processing Summary

Parameters	N
Total cases	7
Excluded cases[(a)]	0
Forecasted cases	0
Newly created cases	0

[(a)] *Cases with a missing value in any variable are excluded from the analysis.*

TABLE 6.43 Variable Processing Summary

		Variables	
		Dependent	**Independent**
		Optimal Bitumen Percentage (OBP)	**Stability (%)**
Number of positive values		7	7
Number of zeros		0	0
Number of negative values		0	0
Number of missing values	User-missing	0	0
	System-missing	0	0

TABLE 6.44 Model Summary and Parameter Estimates

Dependent Variable: Optimal Bitumen Percentage (OBP)								
Equation	**Model Summary**					**Parameter Estimates**		
	R Square	**F**	**df1**	**df2**	**Sig.**	**Constant**	**b1**	**b2**
Quadratic	1.000	45035996273 70560.000	2	4	0.000	0.500	0.005	7.40E-020

Note: The independent variable is stability (%).

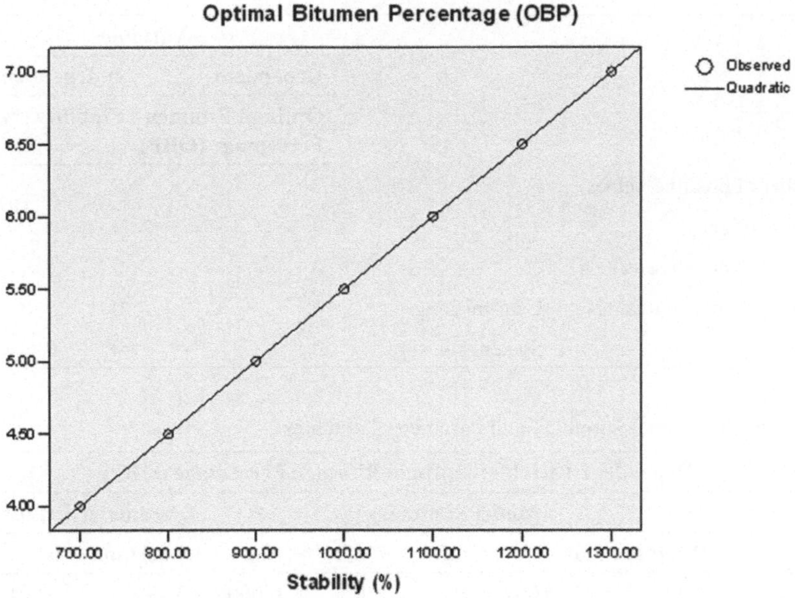

FIGURE 6.16 Scatter diagram for stability and OBP; quadratic model.

TABLE 6.45 Model Description

Model Name		**MOD_2**
Dependent Variable	1	Optimal bitumen percentage (OBP)
Equation	1	Compound[(a)]
Independent variable		Stability (%)
Constant		Included
Variable whose values label observations in plots.		Unspecified

[(a)] *The model requires all non-missing values to be positive.*

TABLE 6.46 Case Processing Summary

	N
Total cases	7
Excluded cases[(a)]	0
Forecasted cases	0
Newly created cases	0

[(a)]*Cases with a missing value in any variable are excluded from the analysis.*

TABLE 6.47 Variable Processing Summary

		Variables	
		Dependent	**Independent**
		Optimal Bitumen Percentage (OBP)	**Stability (%)**
Number of positive values		7	7
Number of zeros		0	0
Number of negative values		0	0
Number of missing values	User-missing	0	0
	System-missing	0	0

TABLE 6.48 Model Summary and Parameter Estimates

Dependent Variable: Optimal Bitumen Percentage (OBP)							
Equation	**Model Summary**					**Parameter Estimates**	
	R Square	**F**	**df1**	**df2**	**Sig.**	**Constant**	**b1**
Compound	0.994	766.057	1	5	0.000	2.139	1.001

Note: The independent variable is stability (%).

TABLE 6.49 Model Description

Model Name		**MOD_3**
Dependent variable	1	Optimal bitumen percentage (OBP)
Equation	1	Growth[(a)]
Independent variable		Stability (%)
Constant		Included
Variable whose values label observations in plots		Unspecified

[(a)]*The model requires all non-missing values to be positive.*

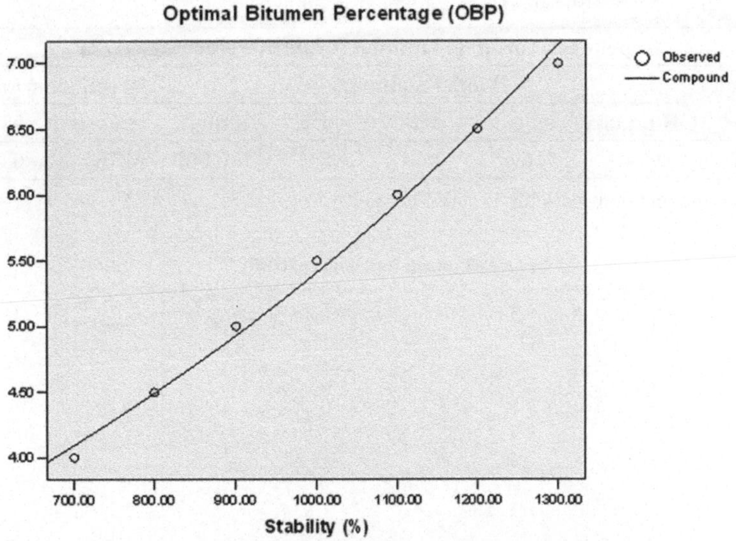

FIGURE 6.17 Scatter diagram for stability and OBP; compound model.

TABLE 6.50 Case Processing Summary

	N
Total cases	7
Excluded cases[(a)]	0
Forecasted cases	0
Newly created cases	0

[(a)]*Cases with a missing value in any variable are excluded from the analysis.*

TABLE 6.51 Variable Processing Summary

		Variables	
		Dependent	**Independent**
		Optimal Bitumen Percentage (OBP)	**Stability (%)**
Number of positive values		7	7
Number of zeros		0	0
Number of negative values		0	0
Number of missing values	User-missing	0	0
	System-missing	0	0

TABLE 6.52 Model Summary and Parameter Estimates

Dependent Variable: Optimal Bitumen Percentage (OBP)							
Equation	**Model Summary**					**Parameter Estimates**	
	R Square	**F**	**df1**	**df2**	**Sig.**	**Constant**	**b1**
Growth	0.994	766.057	1	5	0.000	0.760	0.001

Note: The independent variable is stability (%).

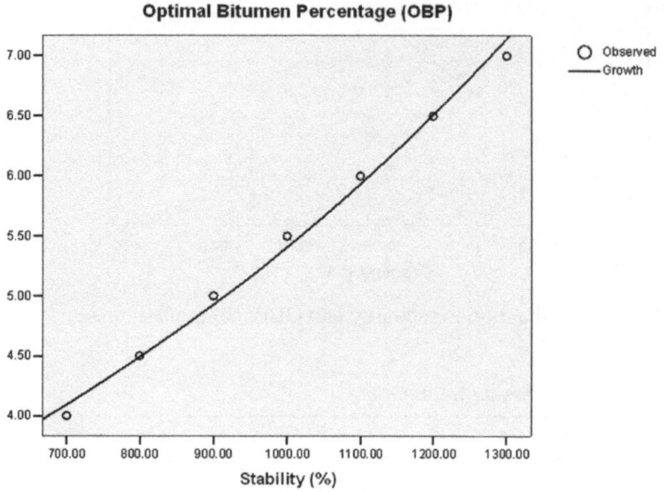

FIGURE 6.18 Scatter diagram for stability and OBP; growth model.

TABLE 6.53 Model Description

Model Name		**MOD_4**
Dependent variable	1	Optimal bitumen percentage (OBP)
Equation	1	Logarithmic
Independent variable		Stability (%)
Constant		Included
Variable whose values label observations in plots.		Unspecified

TABLE 6.54 Case Processing Summary

	N
Total cases	7
Excluded cases[(a)]	0
Forecasted cases	0
Newly created cases	0

[(a)]*Cases with a missing value in any variable are excluded from the analysis.*

TABLE 6.55 Variable Processing Summary

		Variables	
		Dependent	**Independent**
		Optimal Bitumen Percentage (OBP)	**Stability (%)**
Number of positive values		7	7
Number of zeros		0	0
Number of negative values		0	0
Number of missing values	User-missing	0	0
	System-missing	0	0

TABLE 6.56 Model Summary and Parameter Estimates

Dependent Variable: Optimal Bitumen Percentage (OBP)							
Equation	**Model Summary**					**Parameter Estimates**	
	R Square	**F**	**df1**	**df2**	**Sig.**	**Constant**	**b1**
Logarithmic	0.992	625.993	1	5	0.000	–27.844	4.842

Note: The independent variable is stability (%).

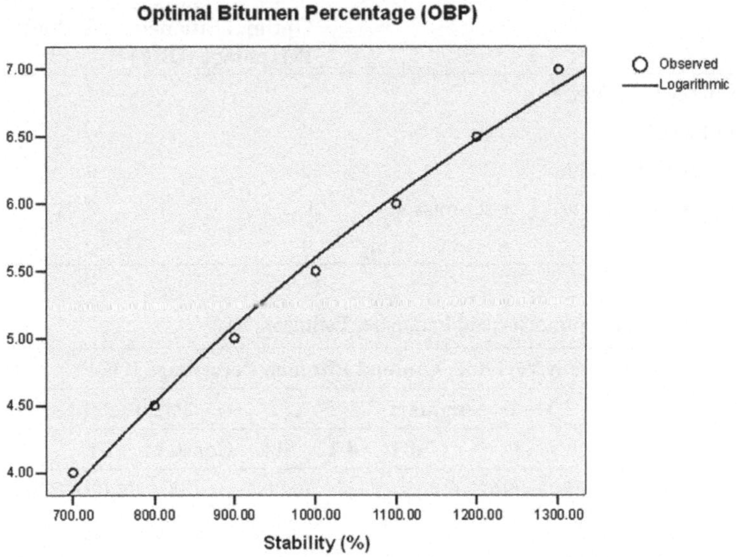

FIGURE 6.19 Scatter diagram for stability and OBP; logarithmic model.

TABLE 6.57 Model Description

Model Name		MOD_5
Dependent variable	1	Optimal bitumen percentage (OBP)
Equation	1	Cubic
Independent variable		Stability (%)
Constant		Included
Variable whose values label observations in plots.		Unspecified
Tolerance for entering terms in equations.		0.0001

TABLE 6.58 Case Processing Summary

Cases	N
Total cases	7
Excluded cases[a]	0
Forecasted cases	0
Newly created cases	0

[a]*Cases with a missing value in any variable are excluded from the analysis.*

TABLE 6.59 Variable Processing Summary

		Variables	
		Dependent	Independent
		Optimal Bitumen Percentage (OBP)	Stability (%)
Number of positive values		7	7
Number of zeros		0	0
Number of negative values		0	0
Number of missing values	User-missing	0	0
	System-missing	0	0

TABLE 6.60 Model Summary and Parameter Estimates

Dependent Variable: Optimal Bitumen Percentage (OBP)									
Equation	Model Summary					Parameter Estimates			
	R Square	F	df1	df2	Sig.	Constant	b1	b2	b3
Cubic	1.000	4503599627370560.000	2	4	0.000	0.500	0.005	7.40E-020	0.000

Note: The independent variable is stability (%).

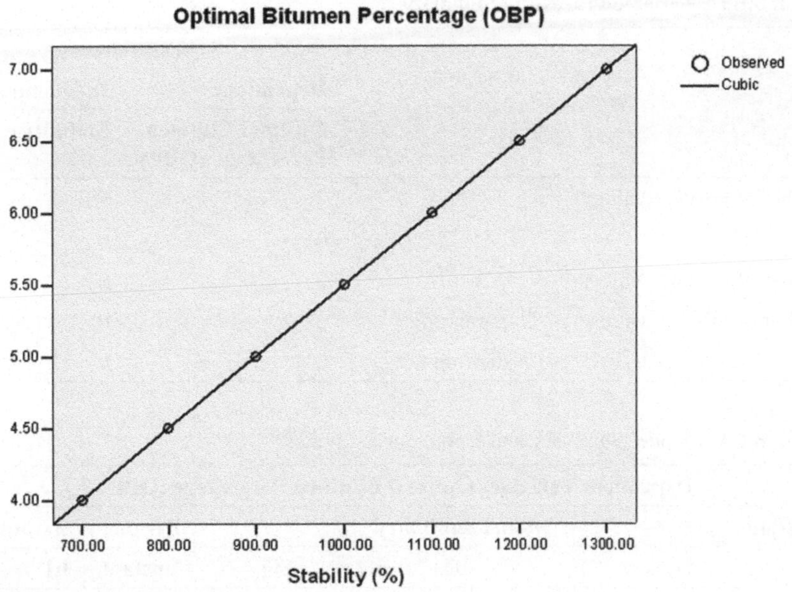

FIGURE 6.20 Scatter diagram for stability and OBP; cubic model.

TABLE 6.61 Model Description

Model Name		**MOD_6**
Dependent variable	1	Optimal bitumen percentage (OBP)
Equation	1	S[(a)]
Independent variable		Stability (%)
Constant		Included
Variable whose values label observations in plots.		Unspecified

[(a)] *The model requires all non-missing values to be positive.*

TABLE 6.62 Case Processing Summary

	N
Total cases	7
Excluded cases[(a)]	0
Forecasted cases	0
Newly created cases	0

[(a)] *Cases with a missing value in any variable are excluded from the analysis.*

TABLE 6.63 Variable Processing Summary

		Variables	
		Dependent	**Independent**
		Optimal Bitumen Percentage (OBP)	**Stability (%)**
Number of positive values		7	7
Number of zeros		0	0
Number of negative values		0	0
Number of missing values	User-missing	0	0
	System-missing	0	0

TABLE 6.64 Model Summary and Parameter Estimates

Dependent Variable: Optimal Bitumen Percentage (OBP)							
Equation	**Model Summary**					**Parameter Estimates**	
	R Square	**F**	**df1**	**df2**	**Sig.**	**Constant**	**b1**
S	0.990	516.419	1	5	0.000	2.571	–846.698

Note: The independent variable is stability (%).

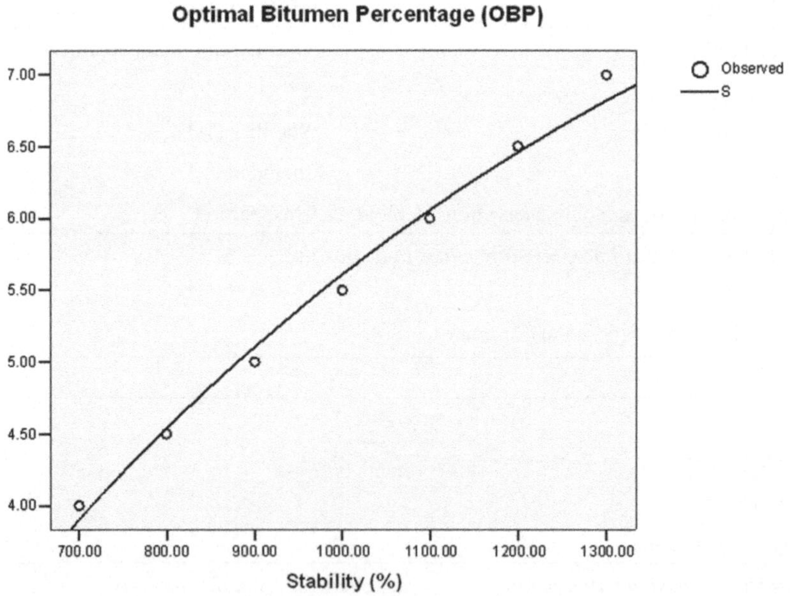

FIGURE 6.21 Scatter diagram for stability and OBP; S model.

TABLE 6.65 Model Description

Model Name		MOD_7
Dependent variable	1	Optimal bitumen percentage (OBP)
Equation	1	Exponential[(a)]
Independent variable		Stability (%)
Constant		Included
Variable whose values label observations in plots.		Unspecified

[(a)]*The model requires all non-missing values to be positive.*

TABLE 6.66 Case Processing Summary

Cases	N
Total cases	7
Excluded cases[(a)]	0
Forecasted cases	0
Newly created cases	0

[(a)]*Cases with a missing value in any variable are excluded from the analysis.*

TABLE 6.67 Variable Processing Summary

		Variables	
		Dependent	**Independent**
		Optimal Bitumen Percentage (OBP)	**Stability (%)**
Number of positive values		7	7
Number of zeros		0	0
Number of negative values		0	0
Number of missing values	User-missing	0	0
	System-missing	0	0

TABLE 6.68 Model Summary and Parameter Estimates

Dependent Variable: Optimal Bitumen Percentage (OBP)							
Equation	**Model Summary**					**Parameter Estimates**	
	R Square	**F**	**df1**	**df2**	**Sig.**	**Constant**	**b1**
Exponential	0.994	766.057	1	5	0.000	2.139	0.001

Note: The independent variable is stability (%).

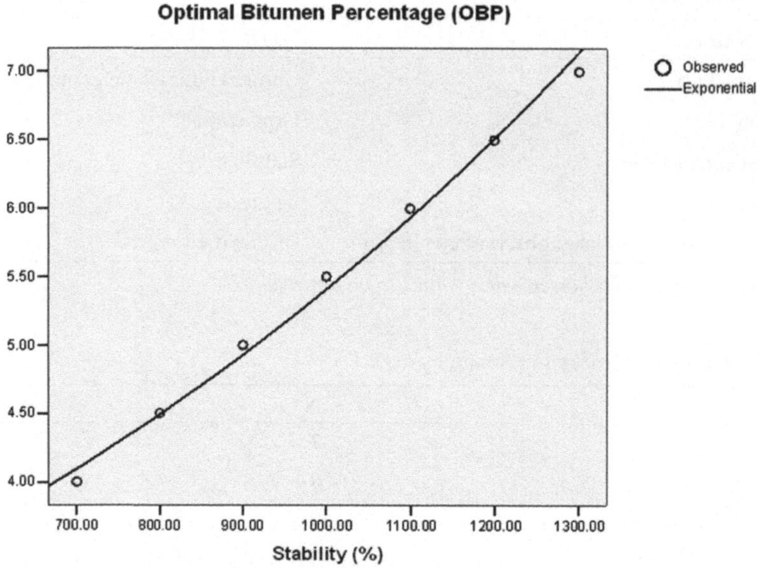

FIGURE 6.22 Scatter diagram for stability and OBP; exponential model.

TABLE 6.69 Model Description

Model Name		MOD_8
Dependent variable	1	Optimal bitumen percentage (OBP)
Equation	1	Inverse
Independent variable		Stability (%)
Constant		Included
Variable whose values label observations in plots.		Unspecified

TABLE 6.70 Case Processing Summary

	N
Total cases	7
Excluded cases[a]	0
Forecasted cases	0
Newly created cases	0

[a] *Cases with a missing value in any variable are excluded from the analysis.*

TABLE 6.71 Variable Processing Summary

		Variables	
		Dependent	Independent
		Optimal Bitumen Percentage (OBP)	Stability (%)
Number of positive values		7	7
Number of zeros		0	0
Number of negative values		0	0
Number of missing values	User-missing	0	0
	System-missing	0	0

TABLE 6.72 Model Summary and Parameter Estimates

Dependent Variable: Optimal Bitumen Percentage (OBP)							
Equation	Model Summary					Parameter Estimates	
	R Square	F	df1	df2	Sig.	Constant	b1
Inverse	0.969	153.805	1	5	0.000	10.193	–4499.674

Note: The independent variable is stability (%).

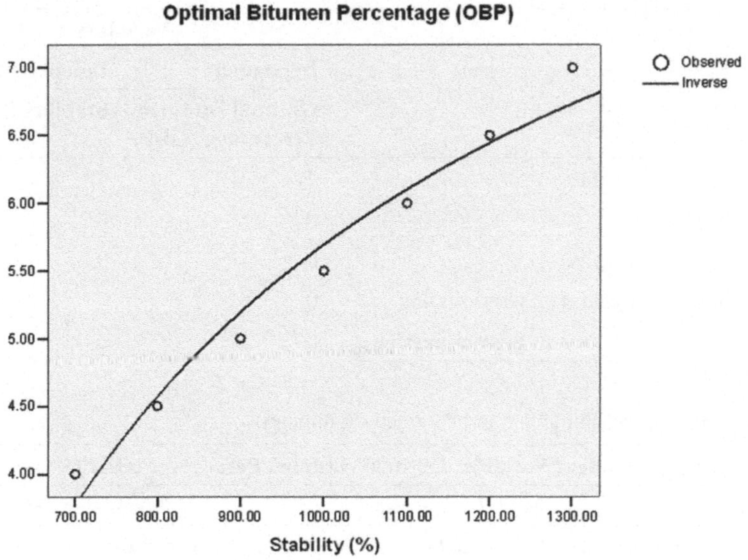

FIGURE 6.23 Scatter diagram for stability and OBP; inverse model.

TABLE 6.73 Model Description

Model Name		MOD_9
Dependent variable	1	Optimal bitumen percentage (OBP)
Equation	1	Power[(a)]
Independent variable		Stability (%)
Constant		Included
Variable whose values label observations in plots		Unspecified

[(a)]*The model requires all non-missing values to be positive.*

TABLE 6.74 Case Processing Summary

	N
Total cases	7
Excluded cases[(a)]	0
Forecasted cases	0
Newly created cases	0

[(a)]*Cases with a missing value in any variable are excluded from the analysis.*

TABLE 6.75 Variable Processing Summary

		Variables	
		Dependent	**Independent**
		Optimal Bitumen Percentage (OBP)	**Stability (%)**
Number of positive values		7	7
Number of zeros		0	0
Number of negative values		0	0
Number of missing values	User-missing	0	0
	System-missing	0	0

TABLE 6.76 Model Summary and Parameter Estimates

Dependent Variable: Optimal Bitumen Percentage (OBP)							
Equation	**Model Summary**					**Parameter Estimates**	
	R Square	**F**	**df1**	**df2**	**Sig.**	**Constant**	**b1**
Power	1.000	68129.882	1	5	0.000	0.011	0.904

Note: The independent variable is stability (%).

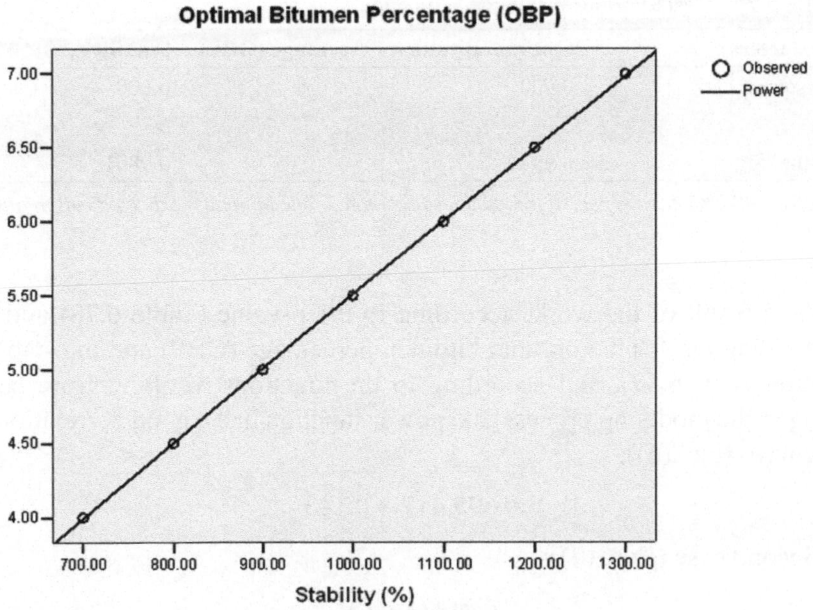

FIGURE 6.24 Scatter diagram for stability and OBP; power model.

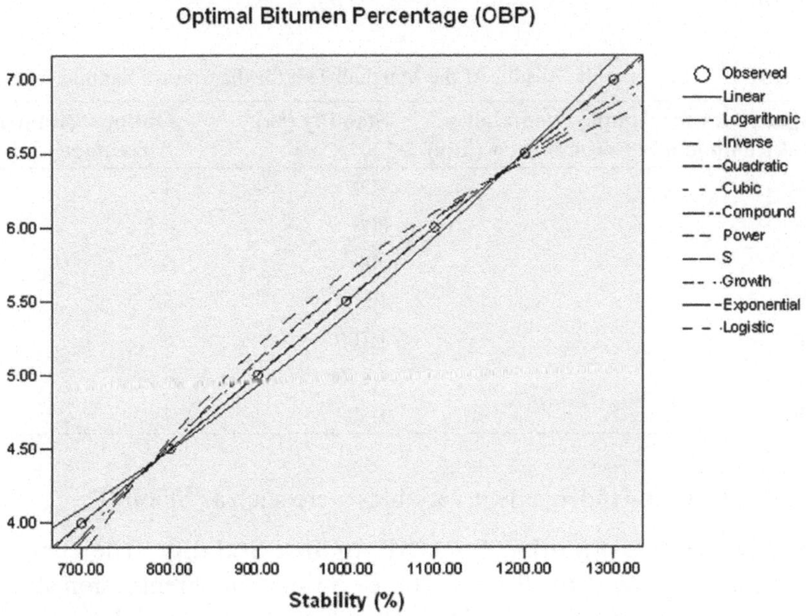

FIGURE 6.25 Regression analysis curve fit for stability (%) as the independent variable and OBP as the dependent variable.

TABLE 6.77 Test Statistics for Logistic Function

Parameters	Optimal Bitumen Percentage (OBP)	Stability (%)
Chi-squar[e(a)]	0.000	0.000
Df	6	6
Asymp. Sig.	0.9972	0.9801

[(a)]7 cells (100.0%) have expected frequencies less than 5. The minimum expected cell frequency is 1.0.

As a result of the work, according to the p-value (Table 6.78) and the scatter diagram for the optimal bitumen percentage (OBP) and the stability (Figures 6.26–6.37), and according to the functions resulting from curve fitting in the modeling process, the power function had a good correlation on the curve (Eqn. (6)):

$$y = 695.417 \times 0.143 \tag{6}$$

Second case (Eqn. (7)):

$$O = f(S, F.I.) \tag{7}$$

where; *S* is the stability of pilot as a field data; F.I. is the fatigue index; and 'O' is the OBP.

TABLE 6.78 Graphs of the Results of the Marshall Test for the Control Sample: Stability

Fatigue Index for Samples Containing 4% of Rhamnolipid Biosurfactant (kPa)	Stability (%)	Optimal Bitumen Percentage (OBP)
1,354	700	4
1,360	800	4.5
1,366	900	5
1,367	1,000	5.5
1,380	1,100	6
1,387	1,200	6.5
1,399	1,300	7

In this work the independent variables were such as Stability:

- **In First Assumption for OBP Against Stability:** The curve estimation procedure allows for quick estimation of regression statistics and the production of related plots for different models (Eqn. (8)).

The nomenclatures for two cases are as the flowing:

- Dependent variable with nomenclature "*o*": OBP; and
- Independent variable with nomenclature "*s*": Stability.

Dependent variable with nomenclature "*O*": OBP, mm, independent variable with nomenclature "*S*" (Eqn. (8)):

$$O = f(S) \tag{8}$$

where; *S* is the stability of pilot as a field data.

The most important effects that were observed based on regression for model summary are as follows, with a focus on the stability of the pilot. The curve fit for dependent and independent variables was analyzed (Figures 6.26–6.37).

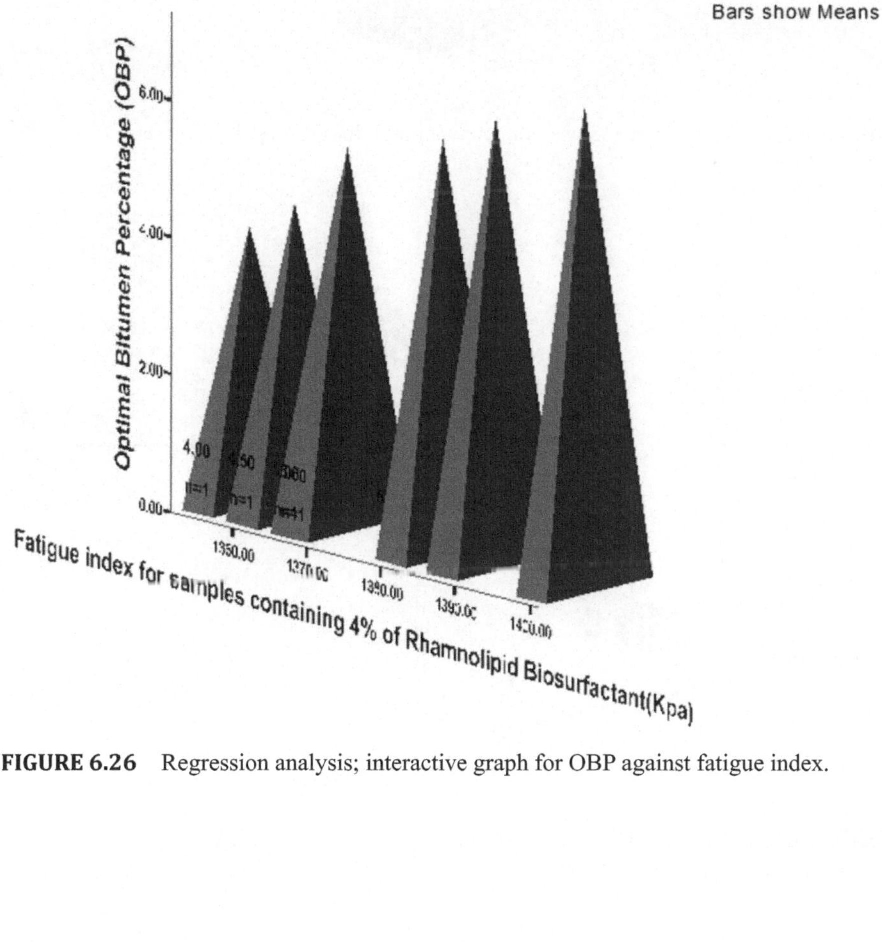

FIGURE 6.26 Regression analysis; interactive graph for OBP against fatigue index.

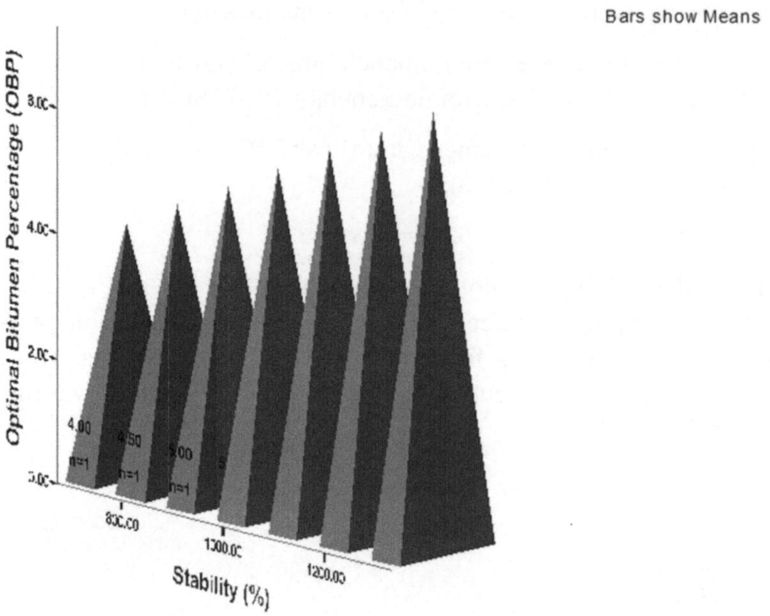

FIGURE 6.27 Regression analysis; Interactive graph for OBP against stability.

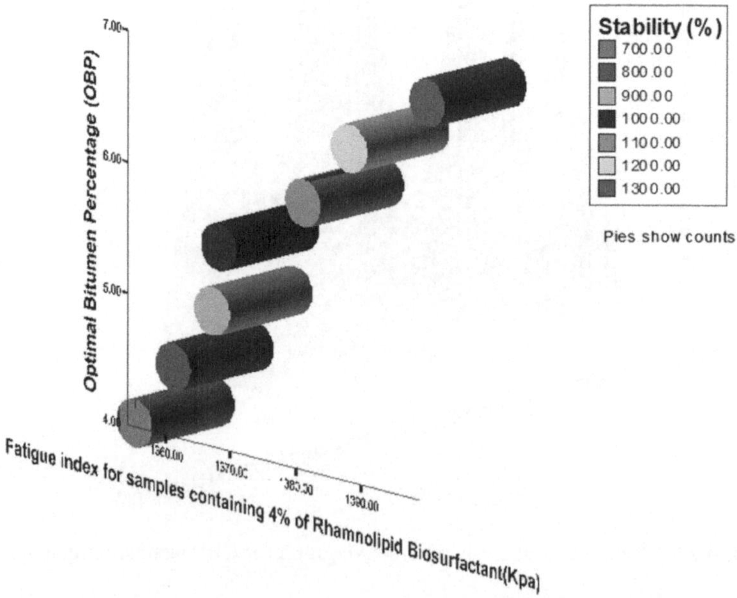

FIGURE 6.28 Regression analysis; interactive graph for OBP against fatigue index.

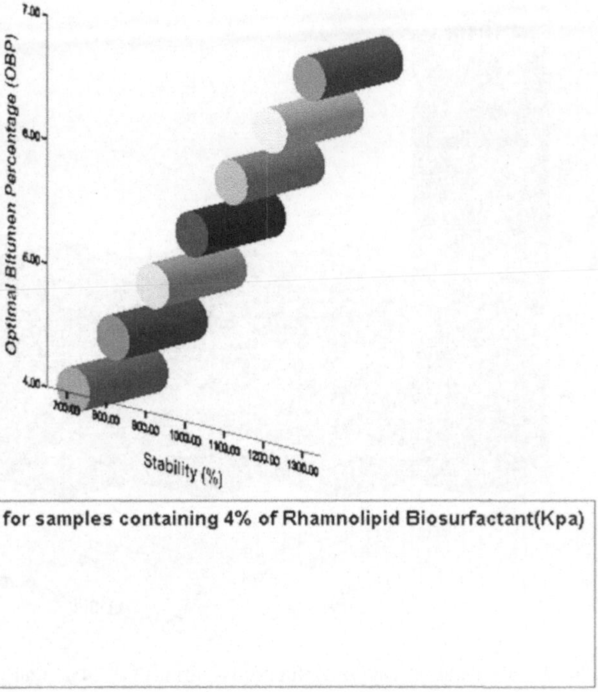

FIGURE 6.29 Regression analysis; interactive graph for OBP against stability.

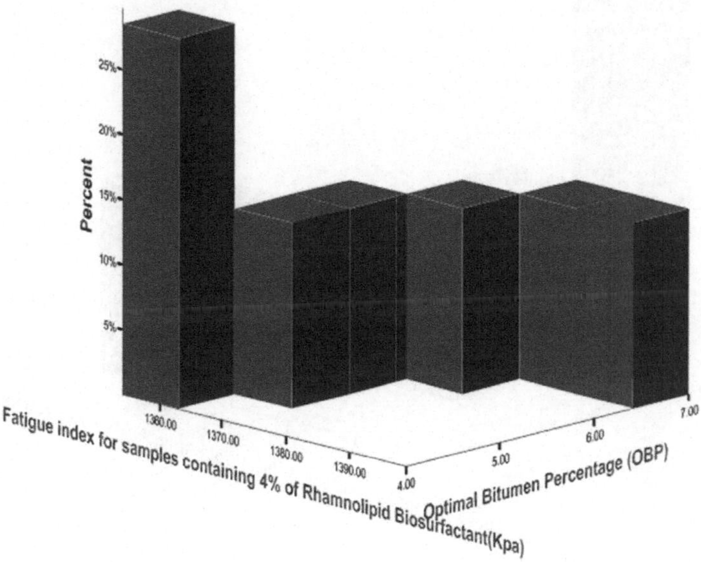

FIGURE 6.30 Regression analysis; interactive graph for OBP against fatigue index.

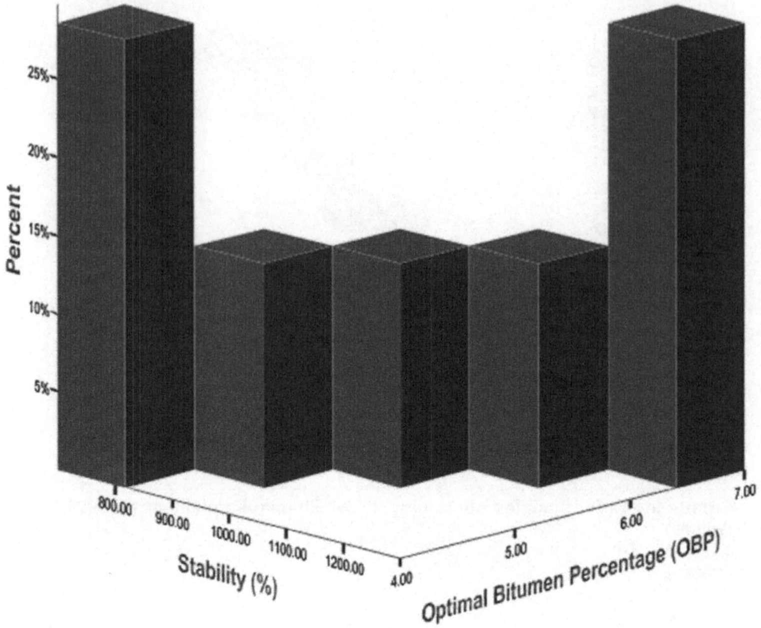

FIGURE 6.31 Regression analysis; interactive graph for OBP against stability.

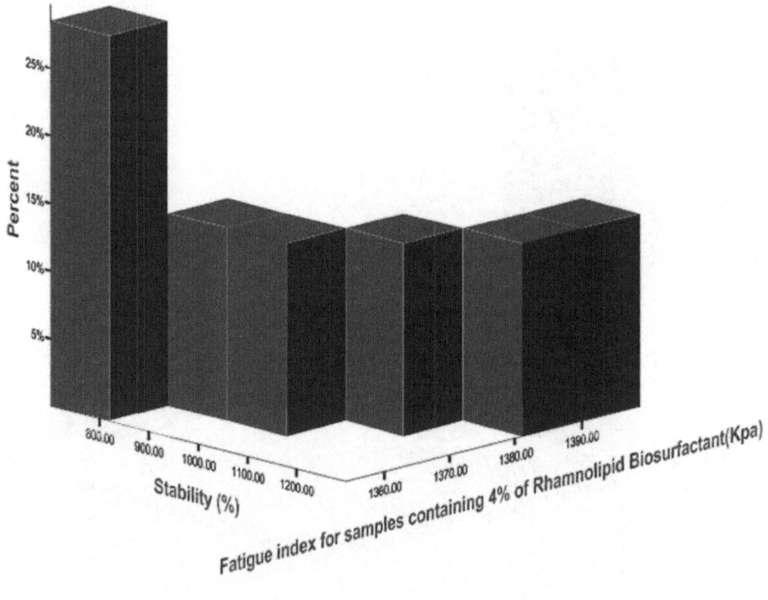

FIGURE 6.32 Regression analysis; interactive graph for fatigue index against stability.

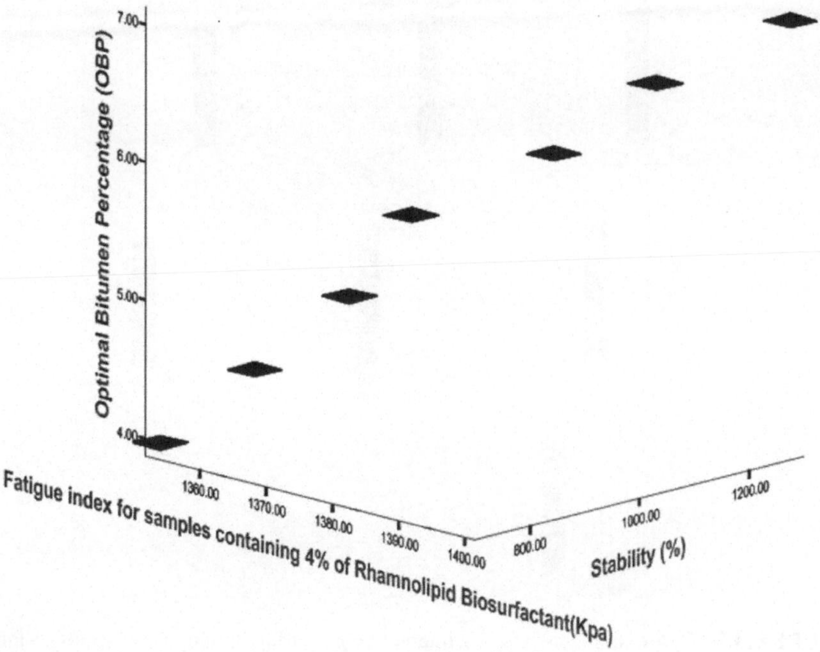

FIGURE 6.33 Regression analysis; interactive graph for OBP against stability and fatigue index.

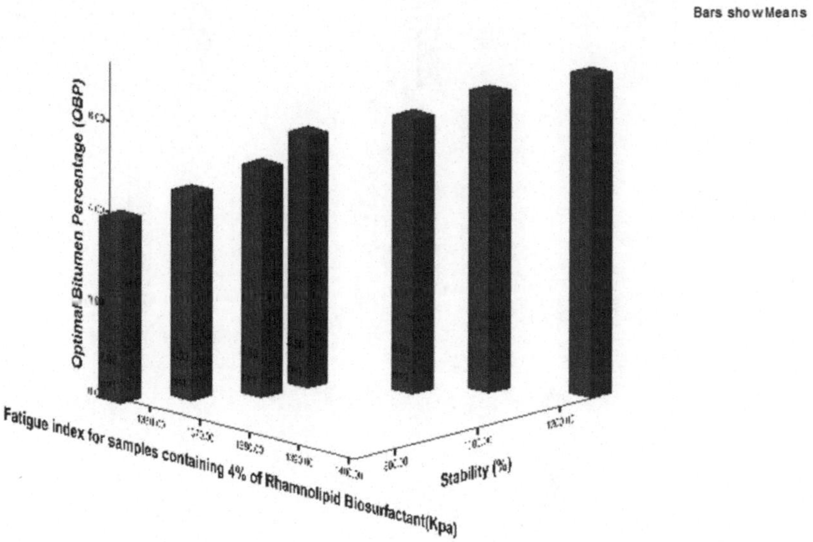

FIGURE 6.34 Regression analysis; 3D interactive graph for OBP against stability and fatigue index.

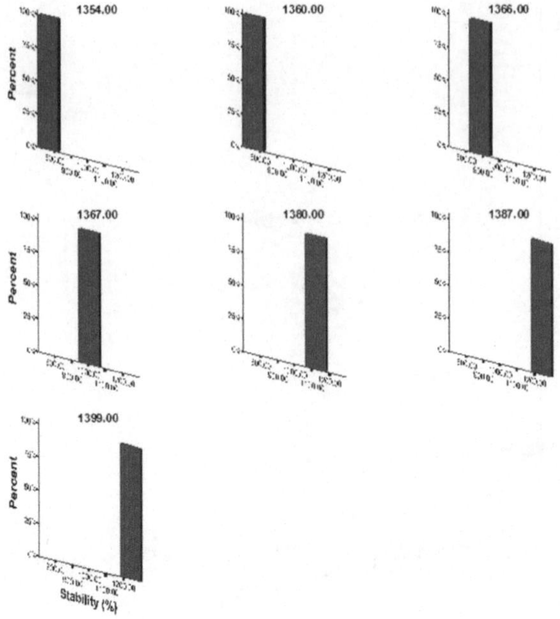

FIGURE 6.35 Regression analysis; 3D interactive graph for stability againstfatigue index.

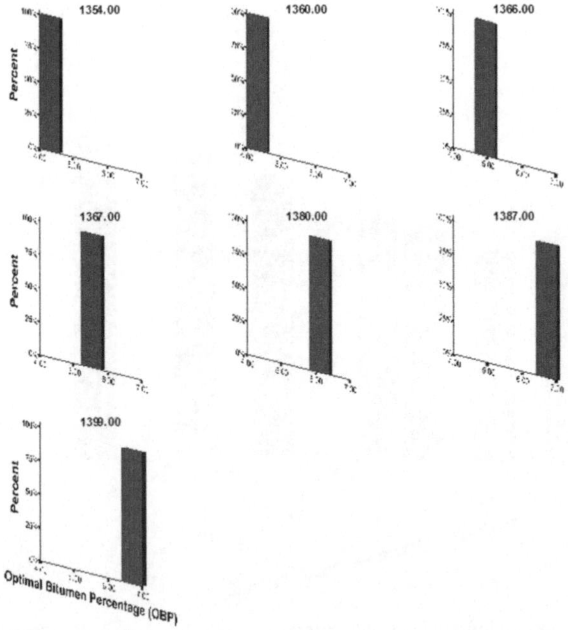

FIGURE 6.36 Regression analysis; 3D interactive graph for OBP against fatigue index.

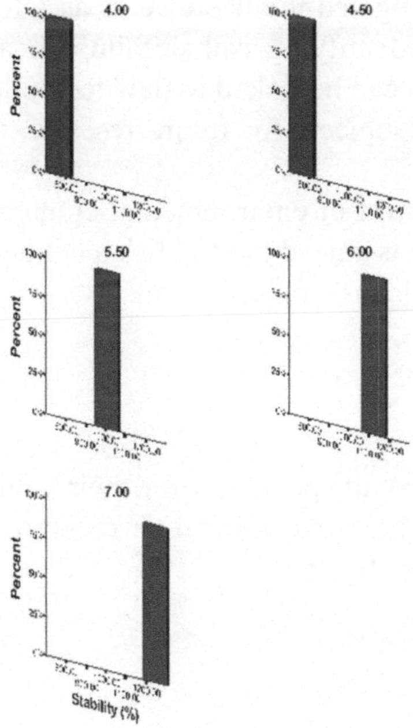

FIGURE 6.37 Regression analysis; 3D interactive graph for stability against OBP.

6.3.4 COMPARISON OF PRESENT WORK WITH OTHER RESEARCH WORKS [65]

This work was compared with a study by Zamanis et al. [65]. The results of this study showed that the connection between GIS, RS, and IoT is a unique technique that is widely used to remove toxic and metallic pollutants from the environment. It also involves promoting the optimal use of Rhamnolipid Biosurfactant.

6.4 CONCLUSION

The main results were as the following:

- The results of the indirect traction fatigue test showed that the sample containing 4% Rhamnolipid Biosurfactant at both 10°C and 20°C has the highest resistance to fatigue.
- The samples containing 4% Rhamnolipid Biosurfactant had the best performance against dynamic loads.

The present work also investigated the use of GIS as a low-cost, high-precision, and rapid method for identifying asphalt failure locations. Finally, this work demonstrated that GIS can be linked to new techniques, including RS and IoT, which are serious subjects for future research in the field of pavement engineering.

This work has led to the control of environmental contaminants caused by toxic agents in asphalt. There is a need to establish a connection between GIS, RS, and IoT for rapid data intercommunication.

ACKNOWLEDGMENT

The author expresses gratitude to all specialists for their valuable observations and advice, as well as to the referees for their recommendations that have enhanced the quality of this chapter.

KEYWORDS

- **fatigue**
- **geospatial information system**
- **internet of things**
- **natural asphalt**
- **networked sensors**
- **regression analysis**
- **remote sensing**
- **rhamnolipid biosurfactant**

REFERENCES

1. Hassan, M., Essam, T., Yassin, A. S., & Salama, A., (2016). Optimization of rhamnolipid production by biodegrading bacterial isolates using Plackett–Burman design. *International Journal of Biological Macromolecules, 82*, 573–579.
2. Read, J., & Whiteoak, D., (2003). *The Shell Bitumen Handbook.* Thomas Telford.
3. Claudy, P., Letoffe, J., King, G., & Plancke, J., (1992). Characterization of asphalts cements by thermomicroscopy and differential scanning calorimetry: Correlation to classic physical properties. *Fuel Science & Technology International.*

4. Masson, J. F., Leblond, V., & Margeson, J., (2006). Bitumen morphologies by phase-detection atomic force microscopy. *Journal of Microscopy, 221*(1), 17–29.
5. Berkers, R. F. A., (2005). Influence of the chemical composition on the machinability of brasses. *Journal of Materials Processing Technology, 170*(1–2), 441–447, ISSN 0924-0136, https://doi.org/10.1016/j.jmatprotec.2005.05.035. https://www.sciencedirect.com/science/article/pii/S092401360500631X.
6. Šaban, M., Porter, S., Costello, C., Djuričić, M., & Vitorović, D., (1980). Polar constituents isolated from Aleksinac oil shale. *Physics and Chemistry of the Earth, 12*, 559–566.
7. Hetényi, M., & Sajgó, C., (1990). Hydrocarbon generation potential of some Hungarian low-rank coals. *Organic Geochemistry, 16*(4–6), 907–916.
8. Longstaff, D., Deo, M., Hanson, F., Oblad, A., & Tsai, C., (1992). Hydrotreating the bitumen-derived hydrocarbon liquid produced in a fluidized-bed pyrolysis reactor. *Fuel, 71*(12), 1407–1419.
9. Binet, S., Pfohl-Leszkowicz, A., Brandt, H., Lafontaine, M., & Castegnaro, M., (2002). Bitumen fumes: Review of work on the potential risk to workers and the present knowledge on its origin. *Science of the Total Environment, 300*(1–3), 37–49.
10. Sato, S., Sugimoto, Y., Sakanishi, K., Saito, I., & Yui, S., (2004). Diesel quality and molecular structure of bitumen-derived middle distillates. *Fuel, 83*(14, 15), 1915–1927.
11. Ji, J., Yao, H., Suo, Z., You, Z., Li, H., Xu, S., & Sun, L., (2016). Effectiveness of vegetable oils as rejuvenators for aged asphalt binders. *Journal of Materials in Civil Engineering, 29*(3), D4016003.
12. Zeng, M., Pan, H., Zhao, Y., & Tian, W., (2016). Evaluation of asphalt binder containing castor oil-based bioasphalt using conventional tests. *Construction and Building Materials, 126*, 537–543.
13. Chen, M., Leng, B., Wu, S., & Sang, Y., (2014). Physical, chemical, and rheological properties of waste edible vegetable oil rejuvenated asphalt binders. *Construction and Building Materials, 66*, 286–298.
14. Safari, A., Ra'ayei, M., Abolhasani, A., & Motamedi, H., (2017). Isolation, screening, and identification of biosurfactant-producing bacteria from the Caspian Sea ecosystem in order to biologically remove environmental pollution. *The First International Conference on Environmental Changes in the Caspian Region*. Mazandaran University, Babolsar.
15. Bagheri, T., Partovi, M., & Bahmani, M., (2013). Production of rhamnolipid biosurfactant by *Pseudomonas aeruginosa* from vegetable oil processing wastes. *New Journal of Cell-Molecular Biotechnology, 2*(9).
16. Kosaric, N., (2001). Biosurfactants and their application for soil bioremediation. *FoodTechnology and Biotechnology, 39*(4), 295–304.
17. Nguyen, T. T., & Sabatini, D. A., (2011). Characterization and emulsification properties of rhamnolipid and sophorolipid biosurfactants and their applications. *International Journal of Molecular Sciences.*
18. Yang, X., & You, Z., (2015). High temperature performance evaluation of bio-oil modified asphalt binders using the DSR and MSCR tests. *Construction and Building Materials, 76*, 380–387.
19. Somé, S. C., Pavoine, A., & Chailleux, E., (2016). Evaluation of the potential use of waste sunflower and rapeseed oils-modified natural bitumen as binders for asphalt pavement design. *International Journal of Pavement Research and Technology, 9*(5), 368.

20. Shirzad, S., Hassan, M. M., Aguirre, M. A., Mohammad, L. N., & Daly, W. H., (2016). Evaluation of sunflower oil as a rejuvenator and its microencapsulation as a healing agent. *Journal of Materials in Civil Engineering, 28*(11), 04016116.
21. Fini, E. H., Kalberer, E. W., Shahbazi, A., Basti, M., You, Z., Ozer, H., et al., (2011). Chemical characterization of biobinder from swine manure: Sustainable modifier for asphalt binder. *Journal of Materials in Civil Engineering, 23*(11), 1506–1513.
22. Asli, H. H., Arabani, M., & Golpour, Y., (2020). Reclaimed asphalt pavement (RAP) based on a geospatial information system (GIS). *Slovak Journal of Civil Engineering* (Vol. 28, No. 2, pp. 36–42). Slovak University of Technology in Bratislava, Slovak. https://doi.org/10.2478/sjce-2020-0013.
23. Azahar, W. N. A. W., Jaya, R. P., Hainin, M. R., Bujang, M., & Ngadi, N., (2016). Chemical modification of waste cooking oil to improve the physical and rheological properties of asphalt binder. *Construction and Building Materials, 126*, 218–226.
24. Ali, A. W., Mehta, Y. A., Nolan, A., Purdy, C., & Bennert, T., (2016). Investigation of the impacts of aging and RAP percentages on effectiveness of asphalt binder rejuvenators. *Construction and Building Materials, 110*, 211–217.
25. Behbahani, H., Ziari, H., Kamboozia, N., Khaki, A. M., & Mirabdolazimi, S., (2015). Evaluation of performance and moisture sensitivity of glasphalt mixtures modified with nanotechnology zycosoil as an anti-stripping additive. *Construction and Building Materials,78*, 60–68.
26. Arabani, M., & Yousefpour, T. M., (2017). Rutting behavior of hot mix asphalt modified by polyvinyl chloride powder. *Petroleum Science and Technology, 35*(15), 1621–1626.
27. Arabani, M., (2011). *Bitumen and Asphalt Tests* (2nd edn.). Gilan University.
28. Zargar, M., Ahmadinia, E., Asli, H., & Karim, M. R., (2012). Investigation of the possibility of using waste cooking oil as a rejuvenating agent for aged bitumen. *Journal of Hazardous Materials, 233*, 254–258.
29. Hariri, A. K., Soltan, A. O., Thomas, S., Deepu, A., & Hariri, A. H., (2017). *Handbook of Research for Fluid & Solid Mechanics*. Published by the Apple Academic Press, Inc., ISBN 9781771885010, USA, Canada. https://doi.org/10.1201/9781315365701 https://www.taylorfrancis.com/books/edit/10.1201/9781315365701/handbook-research-fluid-solid-mechanics-kaveh-hariri-asli-soltan-ali-ogli-aliyev-sabu-thomas-deepu-gopakumar (accessed on 16 January 2023).
30. William, H. K., & Judith, M. T., (1978). Linear hypotheses. *International Encyclopedia of Statistics* (Vol. 1). Free Press.
31. Evan, J. W. (1987). *I. Regression* (pp. 523–541).
32. Julian, C. S. (1967). *II. Analysis of Variance* (pp. 541–554).
33. Asli, H. K., Khodaparast, H. A., Asli, H. H., & Sabermaash, E. E., (2012). Water hammer modeling and simulation by GIS. *Modeling and Simulation in Engineering* (Vol. 1, No. 3, p. 4). Hindawi Publishing Corporation, Article ID 704163. doi:10.1155/2012/704163, ISSN: 16875605, 16875591, Pub., USA. http://www.hindawi.com/journals/mse/aip/704163/ (accessed on 16 January 2023).
34. Lindley, D. V., (1987). Regression and correlation analysis. *New Palgrave: A Dictionary of Economics* (Vol. 4, pp. 120–123).
35. Birkes, D., & Dodge, Y. (1993). *Alternative Methods of Regression*. ISBN 0-471-56881-3.
36. Gardziejczyk, W., (2002). *Texture of Road Surfaces, Methods of Measurement Parameters Evaluation, and its Influence on the Tire/Road Noise* (Issue 6, pp. 23–33). Drogi Mosty, Poland.

37. Content: A Practical Guide, (2007). *Quality Improvement Series* (p. 36–87). 124 National Asphalt Pavement Association, Lanham, MD.
38. Nady, R. M., (1997). *The Quality of Random RAP; Separating Fact from Supposition* (Vol. 2, No. 2, pp. 14–17). National Asphalt Pavement Association, HMAT magazine, Lanham, MD.
39. Newcomb, D. E., Brown, E. R., & Epps, J. A., (2007). Designing HMA Mixtures with High RAP, v(4), pp. 49–57.
40. Kowalski, K. J., McDaniel, R. S., Olek, J., Shah, A., & Li, S., (2008). Development of the international friction index flag value. In: *10th International Conference on Application of Advanced Technologies in Transportation*. Athens, Greece, Paper No. 705.
41. McDaniel, R. S., & Coree, B. J., (2003). *Identification of Laboratory Techniques to Optimize Super pave HMA Surface Friction Characterization*. Phase I: Draft Final Report. Institute for Safe, Quite and Durable Highways, SQDH 2003–6, West Lafayette, Indiana.
42. Li, S., Noureldin, S., & Zhu, K., (2003). *Upgrading the INDOT Pavement Friction Testing Program*. Final Report FHWA/IN/JTRP-2003/23, West Lafayette, Indiana.
43. Asli, H. H., & Arabani, M., (2022). Analysis of strain and failure of asphalt pavement, computational research progress in applied science & engineering. *Transactions of Civil and Environmental Engineering, 8*, 1–11. Article ID: 2250. https://doi.org/10.52547/crpase.8.1.2250.
44. West, T. R., Choi, J. C., Bruner, D. W., Park, H. J., & Cho, K. H., (2001). Evaluation of dolomite and related aggregates used in bituminous overlays for Indiana pavements. *Transportation Research Record, 1757*, 137–147.
45. Indiana Department of Transportation. "INDOT: Districts & Sub-Districts." Indiana Department of Transportation. Archived from the original on 2014-11-07. Retrieved 2011-10-30., Indiana Department of Transportation. Standard Specifications Book. Section 400, Asphalt Pavements, Indianapolis, (2009).
46. Climate.data.org, Climate-Data.org: https://www.yahoo.com/news/weather (accessed on 16 January 2023).
47. Kowalski, K. J., McDaniel, R. S., & Olek, J., (2008). Development of a laboratory procedure to evaluate the influence of aggregate type and mixture proportions on the frictional characteristics of flexible pavements. *Journal of the Association of Asphalt Paving Technologists, Vol. 77*, 35–70.
48. Pezzullo, J. C., (2016). *Handbook of Math Functions by Abramowitz*. Adapted from JavaScript written by PhD, Associate Professor, Pharmacology and Biostatistics Georgetown University Medical Centre, Washington D.C.
49. Recycling Hot Mix Asphalt Pavements. Information Series 123 (IS-123). National Asphalt Pavement Association, Lanham, MD, 27 (2007).
50. Read, J., & Whiteoak, D., (2003). *The Shell Bitumen Handbook*. Thomas Telford.
51. Hariri, A. H., (2022). Investigation of the factors affecting pedestrian accidents in urban roundabouts. *Computational Research Progress in Applied Science & Engineering (CRPASE), 8*(1), 1–4. Article ID: 2255. https://doi.org/10.52547/crpase.8.1.2255. https://crpase.com/archive/CRPASE-Vol-08-issue-01-15172135.pdf (accessed on 16 January 2023).
52. Hariri, A. H., & Nazari, S., (2021). Water age and leakage in reservoirs: Some computational aspects and practical hints. *Larhyss Journal, 48*, 151–167. http://larhyss.net/ojs/index.php/larhyss/article/view/808/807 (accessed on 16 January 2023).

53. Hariri, A. H., & Hozori, A., (2021). Non-revenue water (NRW) and 3D hierarchical model for landslide. *Larhyss Journal, 48*, 189–210. http://larhyss.net/ojs/index.php/larhyss/article/view/810/810 (accessed on 16 January 2023).
54. Safari, A., Rayaee, M., Abolhassani, A., & Motamedi, H., (2008). "Isolation, screening and identification of biosurfactant-producing bacteria from the Caspian Sea ecosystem in order to biologically eliminate environmental pollution;" *The First International Conference on Environmental Change in the Caspian Region*. University of Mazandaran, Babolsar.
55. Kian, H. A., (2024). *Technological Advancement in HVAC&R System Energy Efficiency, Technological Advancement in Clean Energy Production*. Apple Academic Press, Hard ISBN: 9781774915585. https://www.appleacademicpress.com/title.php?id=1416 (accessed on 16 January 2023).
56. Kian, H. A., (2024). IoT-based smart water leak detection system for a sustainable future: A case study. *Sustainable Water Engineering; Smart and Emerging Technologies*. Apple Academic Press, Hard ISBN: 9781774915714. https://appleacademicpress.com/sustainable-water-engineering-smart-and-emerging-technologies/9781774915714 (accessed on 16 January 2023).
57. Amritanshu, S., Kian, H. A., Neha, K. R., Ann, R. A., & Haghi, A. K., (2024). *Technological Advancement in Clean Energy Production*. Apple Academic Press, Hard ISBN: 9781774915585.https://www.appleacademicpress.com/title.php?id=1416 (accessed on 16 January 2023).
58. Ali, P., Kian, H. A., Ann, R. A., & Haghi, A. K., (2024). *Sustainable Water Engineering; Smart and Emerging Technologies*. Apple Academic Press, Hard ISBN: 9781774915714. https://appleacademicpress.com/sustainable-water-engineering-smart-and-emerging-technologies/9781774915714 (accessed on 16 January 2023).
59. Kaveh, H. A., & Kian, H. A., (2022). Isolated pressure zones based on GIS as a solution for water network problems. *Water Practice and Technology*, wpt2022119. doi: https://doi.org/10.2166/wpt.2022.119, https://iwaponline.com/wpt/article/doi/10.2166/wpt.2022.119/ 91338/Isolated-pressure-zones-based-on-GIS-as-a-solution (accessed on 16 January 2023).
60. Hossein, H. A. A. P., Ann, R. A., & Haghi, A. K., (2023). *New Advances in Materials Technologies; Experimental Characterizations, Theoretical Modeling, and Field Practices*. Hard ISBN: 9781774914847. https://www.appleacademicpress.com/new-advances-in-material-technologies-experimental-characterizations-theoretical-modeling-and-field-practices/9781774914847 (accessed on 16 January 2023).
61. Hossein, H. A., Tamara, T., Ann, R. A., & Haghi, A. K. (2023). *Sustainable Water Treatment and Ecosystem Protection Strategies*. Hard ISBN: 9781774915189, https://www.appleacademicpress.com/sustainable-water-treatment-and-ecosystem-protectionstrategies-/9781774915189 (accessed on 16 January 2023).
62. Shahrokhi, M., Feizbahr, M., Choong, K. K., & Rostami, F., (2018). Wave energy dissipation using perforated and non-perforated piles. *International Journal of Engineering Transactions B: Applications, 31*(2), 212–219.
63. Feizbahr, M., Mirhosseini, S. M., & Joshaghani, A. H., (2020). Improving the performance of conventional concrete using multi-walled carbon nanotubes. *Express Nano Letters, 1*(1), 1–9.

64. Akbari, A., Nikookar, M., & Feizbahr, M., (2013). Reviewing performance of piled raft and pile group foundations under the earthquake loads. *Research in Civil and Environmental Engineering, 1*(5), 287–299.
65. Zaumanis, M., Mallick, R. B., Poulikakos, L., & Frank, R., (2014). Influence of six rejuvenators on the performance properties of reclaimed asphalt pavement (RAP) binder and 100% recycled asphalt mixtures. *Construction and Building Materials, 71*, 538–550.

CHAPTER 7

Numerical Simulation Studies of Disasters for Enabling Warning Preparedness and Mitigation: A Case Study

S. S. PRAVEEN and ASWATHY VIJAYAN

Department of Physics, St. Thomas College, Kozhencherry, Pathanamthitta, Kerala, India

ABSTRACT

Post-2004, the preconceived notion about disaster vulnerability changed for the state of Kerala. The subsequent disasters exposed the urgent need for establishing a warning system for various types of natural disasters. Research and development activities are now focused on warning about upcoming disasters and ensuring preparedness and mitigation. Numerical modeling has emerged as a powerful tool and concept for facilitating warnings. This chapter attempts to understand the physics of disasters and the concept of numerical modeling by showcasing a few simulations based on tsunamis and floods along a selected location in Central Kerala, on the Southwest Coast of India. The investigation discusses the inundation and run-up patterns obtained from simulations for different flood and tsunami scenarios.

7.1 INTRODUCTION

The state of Kerala has been affected by several natural disasters. Prior to 2004, the concept of disasters seemed like something out of science fiction books or Hollywood movies, or events that occurred far away. However, the

Sustainability in Energy and Environment: Engineered Materials and Smart Computational Techniques. Ann Rose Abraham, Heru Susanto, A. K. Haghi, & Kian Hariri Asli (Eds.)

tsunami of 2004, the Ockhi cyclone of 2017, the floods of 2018 and 2019, and other associated disasters shattered these preconceived notions. In the aftermath of each disaster, research and development efforts gained momentum, leading to the establishment of a tsunami warning center in India, a cyclone warning center in Thiruvananthapuram, and various others. Currently, Kerala is actively engaged in research activities for the implementation of a flood warning unit. Regardless of the type of disaster, a crucial component of any warning system is the use of prediction models, which can only be achieved through numerical modeling. Numerical modeling is an excellent tool for analyzing past events and predicting future ones. A real-time disaster can occur at any moment and can be caused by various combinations of factors. Therefore, when predicting future scenarios, it is crucial to consider all these combinations, where numerical modeling can play a significant role. Among many disasters, the impact of tsunamis and floods was particularly severe in Kerala. At this point, this chapter investigates these disasters, their physics, modeling concepts, and conducts a simulation to compute inundation and run-up for a selected region in Central Kerala.

7.2 BASICS ABOUT TSUNAMI

Tsunamis are series of water waves generated by vertical displacement of the ocean surface and are usually triggered by seismic activity, undersea volcanic eruptions, submarine landslides, nuclear explosions, or meteorite impacts within the ocean. The tsunami waves initially triggered in the ocean will travel into rivers, estuaries, coastal bays, and gulfs. The propagation of a tsunami is dependent on the depth of water. The term tsunami comes from the Japanese words for harbor wave (harbor (tsu) and wave (nami)). Tsunami waves are notorious for generating resonant phenomena in harbors after tsunamigenic offshore earthquakes.

The main characteristic of a tsunami wave is that it is different from wind waves. Tsunamis are shallow water waves involving long periods and wavelengths. When tsunami waves have wavelengths of kilometers, wind waves have shorter wavelengths of meters and periods of about 10 s. The wavelength of a tsunami in the deep ocean is around 100 km and it has a period of one hour. Tsunamis are called shallow water waves because the ratio between the water depth (d) and its wavelength (L) is very small, i.e., $d/L < 0.05$. The velocity of a tsunami wave is computed using the equation $c = \sqrt{gd}$, where $g = 9.8$ m/s^2 is the acceleration due to gravity and d is the depth of the ocean. The transformation of a tsunami occurs as it leaves the deep ocean and enters shallow water, where the modification of tsunami characteristics begins as shown in Figure 7.1.

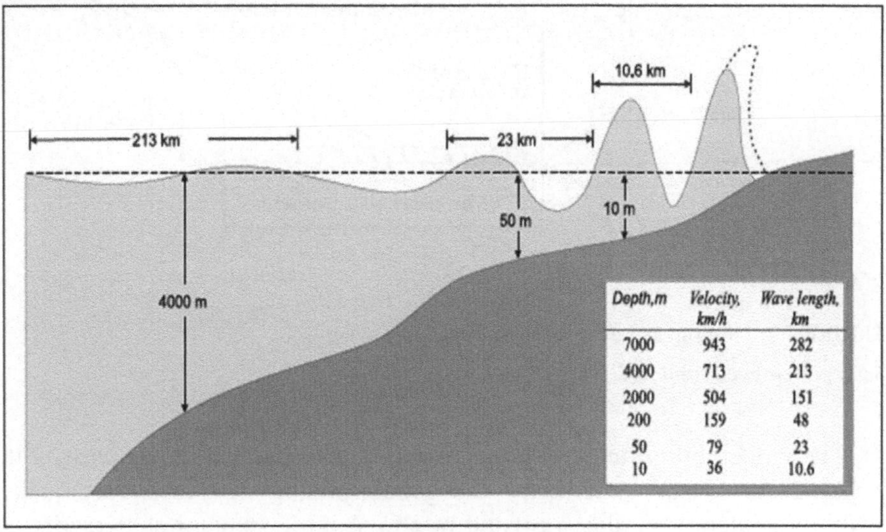

FIGURE 7.1 Propagation characteristics of a tsunami wave.
Source: Reprinted from Ref. [1]

The speed of a tsunami decreases as the water depth decreases while maintaining a constant energy flux. When tsunami waves approach the nearshore area, they interact with the ocean floor, causing the wavelength to compress and the height to increase to maximum values depending on the characteristics of the source and the bathymetric features. Tsunamis have the potential to cause significant damage by eroding beaches, vegetation, trees, and causing severe flooding that can demolish coastal areas and structures.

7.2.1 TSUNAMI RUN-UP AND INUNDATION

The most pertinent characteristics of tsunami waves are their inundation and run-up. Inundation refers to the horizontal extent of water that enters the land and is usually measured with reference to the shoreline. Run-up, on the other hand, refers to the vertical extent of water that rises above the mean sea level (MSL) and is measured with reference to it (Figure 7.2).

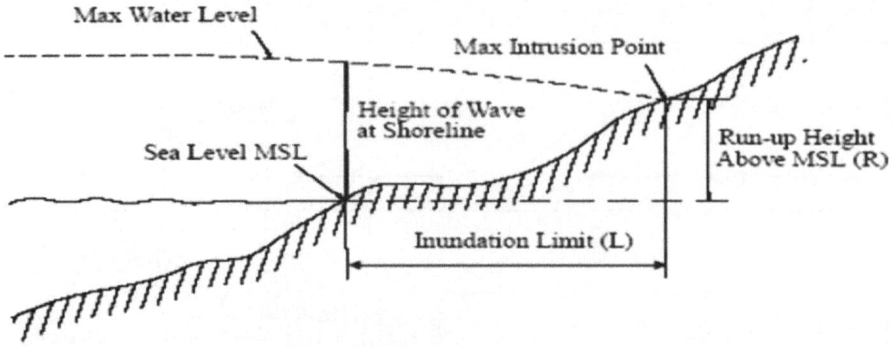

FIGURE 7.2 Run-up and inundation of tsunami.
Source: Reprinted from Ref. [1]

The vulnerability range in terms of run-up and inundation can vary with reference to coastal bathymetry and geomorphology. Besides, earthquake fault parameters also play a pivotal role in shaping tsunami characteristics along the coast. Earthquake fault parameters are the triggering factors of an earthquake which, in turn, can lead to a tsunami (Table 7.1).

TABLE 7.1 Seismic Parameters in Respect of 1945 Makran and 2004 Sumatra Earthquakes*

SL. No.	Source Parameter	1945 Makran	2004 Sumatra
1.	Longitude (degrees)	63 E	95.85 E
2.	Latitude (degrees)	24.5 N	3.32 N
3.	Magnitude (M_w)	8.0	9.3
4.	Slip (m)	15	15
5.	Fault length (km)	200	680
6.	Fault width (km)	100	150
7.	Strike (degree)	270	342
8.	Dip (degree)	15	15
9.	Rake (degree)	90	90
10.	Depth (km)	30	20

**Some values were adapted and organized using Praveen's PhD Thesis (2012 – No permission needed).*

7.3 ABOUT FLOODS IN KERALA

Kerala has experienced several floods throughout its history, notably in 1907, 1924, 1960, 2018, and 2019. However, it was the floods of 2018 and their recurrence in 2019 that revealed the lack of preparedness and awareness regarding such disasters. The floods of 1907, occurring in July and August, recorded a staggering 1,780 mm of rainfall, which was 175% higher than the average rainfall in Kerala. The floods of 1924, known as a catastrophic event, took place in July. The sudden release of water from the Periyar dam, combined with heavy rainfall, resulted in this devastating flood, with a total rainfall of approximately 3,368 mm. The floods of 1960, spanning from June 4th to August 21st, received 2,387 mm of rainfall, which was 50% above the normal rainfall levels. The floods of 2018 occurred in the month of August, mainly due to external dam discharge, improper land cover, and land use activities. During this time, Kerala received 23,466 mm of rainfall between June 1 and August 20. Among the total 14 districts, seven experienced severe flooding. However, the floods of 2019 were comparatively less severe than the 2018 floods, which occurred between August 8 and August 29, as there were no dam openings or any other external discharges like in the 2018 event.

7.4 FLOOD AND ITS MAJOR CAUSES

Flood is one of the most dangerous natural hazards and is a worldwide phenomenon. When a large body of water from the sea, dams, rivers, canals, or drainages overflows into normally dry land due to heavy rainfall, it is called a flood. Flooding is a temporary natural phenomenon. Floods are devastating hazards in terms of property damage and human casualties, among other things. Some major causes of floods are:

1. **Heavy Rain:** Higher precipitation due to cyclones and depressions formed at sea can result in increased runoff. However, sometimes the water does not drain properly into the basins or reservoirs due to the construction of large buildings, blockages, and when heavy rain persists over a long period of time, the water level rises into homes.
2. **Overflowing Rivers:** It occurs when we experience heavy rain and leads to the river overflowing due to resistance offered by the bottom of the riverbed and the lack of water holding capacity of the soil. The heavy rainfall causes the river to be unable to carry the increased

volume of water, resulting in over-bank flow that causes the neighboring land area to be inundated.

3. **Broken Dams:** During heavy rains, the dams fail to hold too much water due to their old infrastructure. The water exerts pressure on the walls of the dams, leading to the bursting of dams due to the lack of water carrying capacity. Sometimes, excess water can be released from the dams to prevent them from breaking, which can also cause severe floods.
4. **Urban Drainage Basins:** Many of our cities are composed of concrete and other impermeable materials, which prevent water from seeping into the ground. This leads to flooding in low-lying areas within urban cities.
5. **Lack of Vegetation:** Vegetation is the most common factor that helps to slow down runoff and prevent floods. Trees and plants have the ability to absorb water. However, due to human activities such as deforestation, forests are being cut down or burned.

7.4.1 TYPES OF FLOODS

Floods are categorized based on size, area of extent, duration, speed, geography, or cause of flooding. Based on these factors, floods can be classified as flash floods, coastal floods, urban floods, river floods, and ponding. Among these, flash floods are the most dangerous and devastating. Floods can be divided into four types: riverine floods, coastal floods, urban floods, and flash floods.

1. **Riverine Flood:** Heavy rainfall leads to a decrease in the water holding capacity of the soil in the river, which in turn affects the capacity of the river. When the river's capacity is exceeded, the riverbank overflows and flooding occurs along the river basin. This type of flooding is typically a slow process, taking several days to cause a disaster.
2. **Coastal Flood:** It is mainly due to cyclonic activities, such as tropical cyclones, that heavy rainfall occurs. Sometimes, earthquakes and volcanoes can cause large tidal waves, resulting in the flow of seawater along the coastal area.
3. **Urban Flood:** This type of flood is mainly due to the blockage of the drainage system, which decreases the efficiency of the water carrying capacity of the system and also affects its proper functioning. As a

result, the water in the drainage system cannot flow freely into the river system. Another reason for urban flooding is the presence of concrete surfaces, which reduce water infiltration into the ground and increase runoff, leading to floods.

4. **Flash Flood:** Flash floods are the most dangerous type of floods. They occur within six hours of heavy rainfall and can cause significant damage. Flash floods are characterized by short, intense rainfall rates. Sometimes, heavy rainfall can occur suddenly within a short period of time, resulting in the rapid movement of water across land areas and causing severe flooding. Flash floods can be triggered by events such as dam breaks or levee failures, making flood-prone areas more vulnerable. There are two types of flash floods: natural and artificial. Natural flash floods occur when heavy rainfall affects low-lying areas, causing water to flow downstream. Artificial flash floods, on the other hand, are caused by the sudden release of water from dams due to dam failures. Flash floods are unpredictable and challenging to manage in real time.

7.4.1.1 PHYSICS OF TSUNAMI MODELLING

The TUNAMIN2 model is widely accepted and globally recognized for estimating and simulating tsunami characteristics [1–6]. The concept of tsunami modeling originates from long wave theory, specifically in very shallow water where the ratio of water depth to wavelength is very small. The TUNAMIN2 model incorporates the following features:

- The effect of bottom friction during propagation is very significant in computing attenuation [7].
- The model is capable of simulating reflection for fixed coastline boundaries and open boundaries.
- The depth of the ocean is considered a relevant parameter for calculating inundation and run-up.
- The horizontal eddy turbulence is considered negligible when compared to bottom friction, except for run-up on land.
- The TUNAMIN2 model is exclusively designed for tsunamis, and does not consider any other wave formations.
- The model has options for incorporating tidal fluctuations.

The basic equations associated with this model are given below:

$$\frac{\partial \eta}{\partial t}+\frac{\partial M}{\partial x}+\frac{\partial N}{\partial y}=0$$

$$\frac{\partial M}{\partial t}+\frac{\partial}{\partial x}\frac{(M^2)}{(D)}+\frac{\partial}{\partial x}\frac{(MN)}{(D)}+gD\frac{\partial \eta}{\partial x}+\frac{gn^2}{D^{7/3}}M\sqrt{M^2+N^2}=0$$

$$\frac{\partial N}{\partial t}+\frac{\partial}{\partial x}\frac{(MN)}{(D)}+\frac{\partial}{\partial y}\frac{(N^2)}{(D)}+gD\frac{\partial \eta}{\partial x}+\frac{gn^2}{D^{7/3}}N\sqrt{M^2+N^2}=0$$

where M and N are the discharge fluxes in the x and y directions, u and v are the vertically averaged horizontal particle velocities, t is the time, g is the gravitational acceleration, h is still water depth, η is the vertical displacement of the water surface above the still water level ($z = 0$), D is the total water depth ($h + \eta$), and n is Manning's roughness coefficient. The bottom frictions in the x and y directions are expressed as:

$$M=\int_{-h}^{\eta} u\,dz=u(h+\eta)=uD$$

$$N=\int_{-h}^{v} v\,dz=v(h+\eta)=vD$$

$$\frac{\tau_x}{\rho}=\frac{fn^2}{D^{7/3}}M\sqrt{M^2+N^2}$$

$$\frac{\tau_y}{\rho}=\frac{fn^2}{D^{3/2}}N\sqrt{M^2+N^2}$$

where f is the friction coefficient. The nonlinear terms are kept only for the computations performed in the inner and higher resolution grid, namely Grid D (Figure 7.3).

The above set of equations were solved by finite difference method [2]. The leapfrog scheme, which introduces a second-order truncation error, is utilized. The Courant, Friedrichs, and Lewy (CFL) criteria are employed to assess the stability of the model. The model grid cells are classified as either dry or submerged based on the condition of the total water depth ($D = h + \eta > 0$ indicates a submerged cell, while $D = h + \eta < 0$ indicates a dry cell). The wave front is located using this classification, and discharge fluxes are computed accordingly. The model and its governing equations are described in detail by various authors, including Kajiura [8], Shuto & Goto [9], Takeda [10], Hibbard & Peregrine [11], Aida [12], Houston & Butler [13], Goto & Shuto [14], Uda [15], Kajiura [16], Mansinha & Smylie [17], Okada [18], and Imamura [19].

7.4.1.2 STUDY AREA FOR TSUNAMI MODELLING

Ernakulum District in Central Kerala has been chosen as the study area for tsunami model computations. The coastal zone of Ernakulam district is divided into two sectors: (i) Vayalar-Manasseri; and (ii) Manasseri-Munambam (Figure 7.4). These two areas were selected for model computations due to their low-lying nature in Central Kerala. The simulation was conducted for three sources of tsunami generation: Sumatra 2004, Makran 1945, and a hypothetical worst-case scenario.

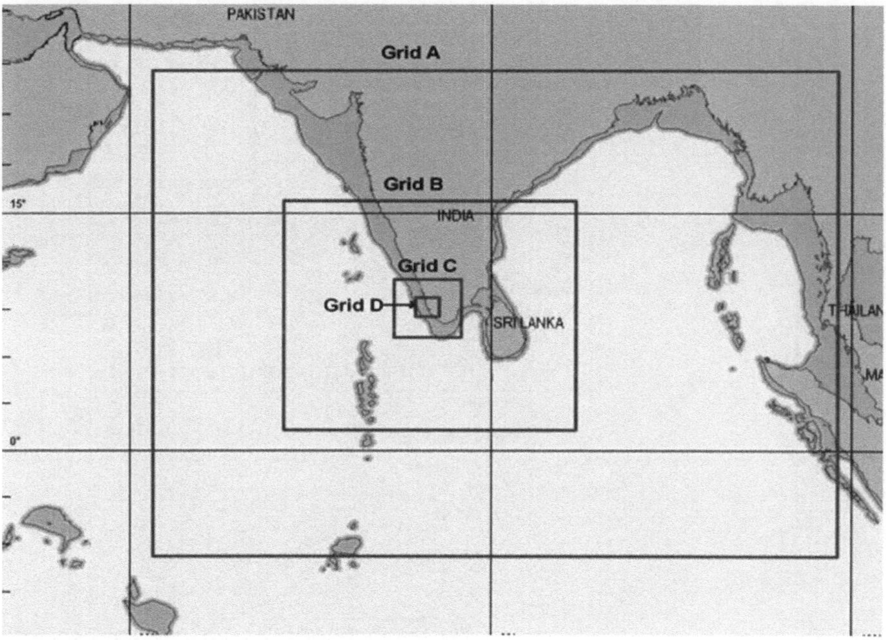

FIGURE 7.3 Grid pattern of tsunami modelling.

Source: Reprinted from Ref. [20]

7.5 PHYSICS OF FLOOD MODELLING

The propagation of a flood is in the form of a wave that propagates from upstream to downstream, as shown in Figure 7.5. This wave is associated with varying parameters such as velocity, depth, displacement, time, and distance. The relevant topics of flood modeling, the equations behind flood modeling, and the concepts about the flood warning system are explained below:

$$\text{Velocity} = \frac{\textit{Displacement}}{\textit{Time}}$$

$$V = \frac{D}{T} \quad (1)$$

$$D = V \times T$$

$$D = V_F \times \Delta t \quad (2)$$

where; V_F is the flood velocity; 't' is the propagation time; and 'D' is the displacement.

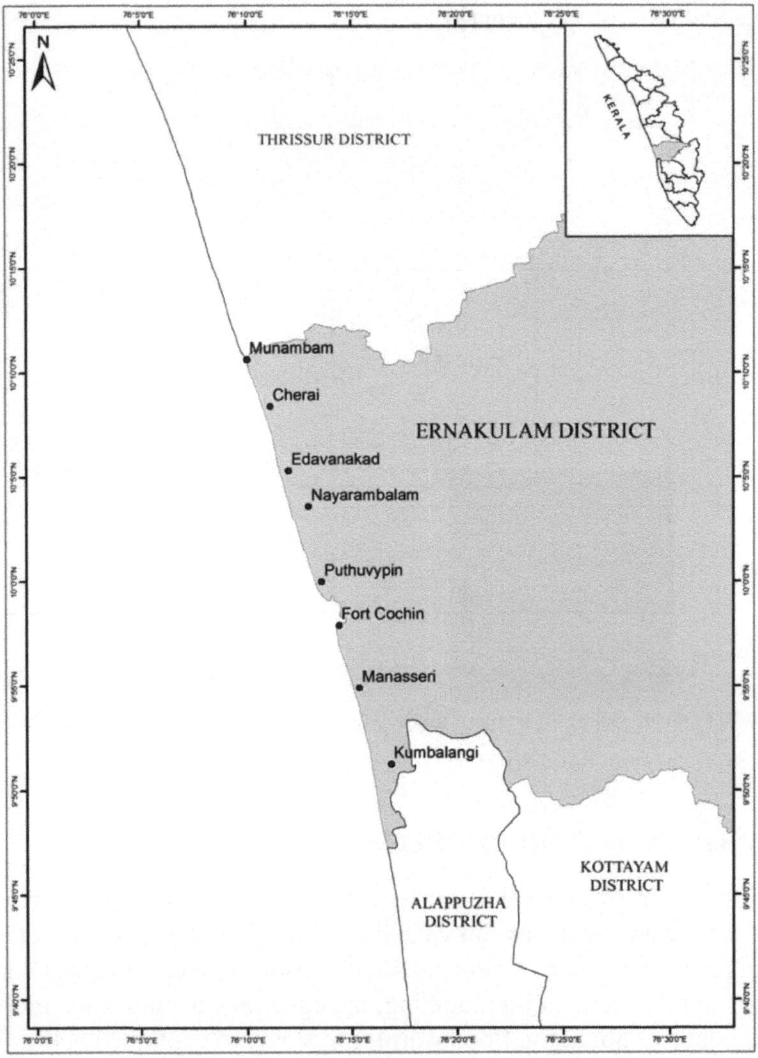

FIGURE 7.4 Location map of Ernakulam district.

Source: Reprinted from Ref. [1]

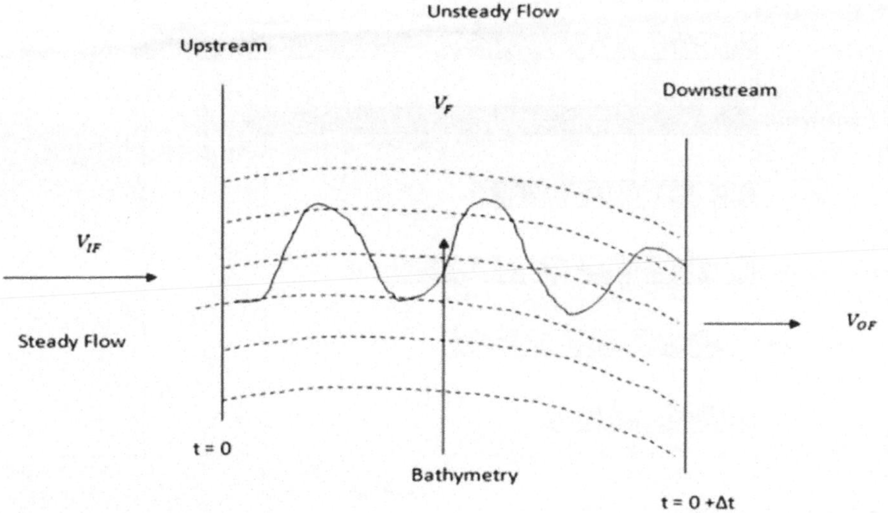

FIGURE 7.5 Propagation of a flood wave.
Source: Reprinted from Ref. [20].

7.5.1 STUDY AREA

The study area is the Periyar river basin, which lies between a north latitude of 9°15'30" and 10°21'00" and an east longitude of 76°08'38" and 77°24'32." The river flows from the Western Ghats to the Lakshadweep Sea. It is known as the lifeline of Kerala because it is the backbone of Kerala's economy. The Periyar is the longest river in Kerala, with a length of 244 km. It is also the second largest river basin in Kerala. Near Aluva, it divides into two distributaries: the MarthandaVarma distributary and the Mangalappuzha. The entire Periyar network system is controlled by 17 dams and 2 barrages. The entire Periyar network system is controlled by 17 dams and 2 barrages. The historical analysis of rainfall statistics in Kerala is shown in Figure 7.6.

Periyar drains through Idukki, Ernakulam, and Thrissur districts in Kerala and Coimbatore district in Tamil Nadu. It originates from the Sundara Mala in Tamil Nadu. The total catchment area of the river is 5,398 sq km. Its 5,284 sq km lies in Kerala, and the remaining 114 sq km lies in Tamil Nadu. The major tributaries are Mullayar River, Cheruthoni River, Muthirapuzha, Perinjankutty River, and Idamalayar.

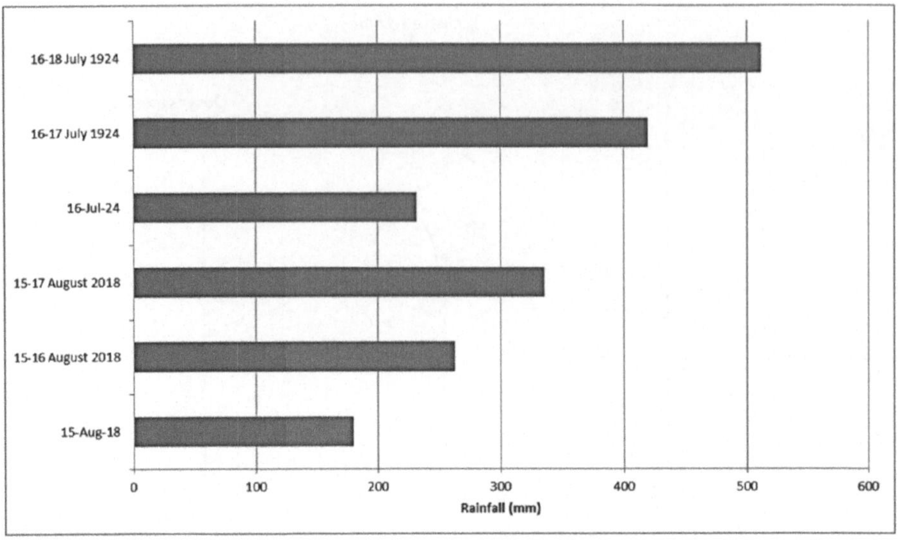

FIGURE 7.6 Report of Central Water Commission-Study Report of Kerala Floods. 2018.

7.6 MODEL FOR TSUNAMI

Several models are available for tsunami inundation modeling. Basically, all of these models primarily address tsunami generation, propagation, and coastal inundation aspects. Some of the most important global tsunami models include: (i) MOST model; (ii) TOAST model; (iii) Delft model; (iv) Italy model; (v) ECJRC model; (vi) ADCIRC; (vii) COULWAVE; and (viii) TUNAMIN2.

The mathematical code for the TUNAMIN2 model was originally authored by Professors Nobuo Shuto and Fumihiko Imamura of the Disaster Control Research Center at Tohoku University in Japan, as part of the tsunami inundation modeling exchange (TIME) program [1, 21]. TUNAMIN2 stands for "Tohoku University's Numerical Analysis Model for Investigation of Near-field tsunamis, No. 2."

TUNAMIN2 provides relevant numerical tools for the propagation and coastal amplification of tsunamis under different initial conditions [4]. It solves the linear form of the long-wave equations in spherical coordinates in the largest domain (the deep sea), and the nonlinear shallow-water equations in Cartesian coordinates in smaller domains with finer grids for basins of irregular shape and topography. The equations are solved using the leap-frog scheme of finite differences [21]. The selection of the TUNAMIN2 model is primarily based on its inputs, which can be classified into three categories: (i) a source that reflects the known geology and seismology of the event; (ii) ocean bathymetry and coastal topography; and (iii) a tsunami propagation and run-up model.

From the reviews of various literature carried out under this investigation, it was found that only two models, namely TUNAMIN2 and MOST, are capable of modeling the process of a tsunami starting from its generation offshore to its inundation at the coast. The common feature of both these models is that they use finite difference mathematical algorithms to solve shallow water equations in a nonlinear mode. MOST uses a variable computational grid, no friction factors, and its run-up prediction is based on the elevation of the last wet grid point the wave encounters as it propagates up on the beach. On the other hand, TUNAMIN2 uses a fixed computational grid, involves a friction factor, and allows the wave to advance past the initial shoreline. Moreover, its maximum run-up prediction is based on the maximum wave height at the shoreline. Additionally, TUNAMIN2 gives more prominence to fine grid data, which is of utmost importance when dealing with the run-up and inundation characteristics of the coast. Furthermore, TUNAMIN2 is more widely and globally accepted for the simulation of tsunami inundation. The Indian Tsunami Warning Center at the Indian National Center for Ocean Information Services (INCOIS) also uses this model for mathematical computations, warnings, and advisories [22].

7.6.1 MODEL FOR FLOOD

HEC-RAS is the most commonly used open-source computer program that helps to model the hydraulics of water flow through rivers and other channels. This integrated program was developed by the US Department of Defense, Army Corps of Engineers and was developed in 1995. The HEC-RAS model is used for planning, flood risk reduction, environmental impact assessment and mitigation, and restoration. HEC-RAS is an integrated system software that is essentially a multitasking and multiuser network. It is used for both steady and unsteady flow hydraulics, providing higher data storage, management, graphics, and reporting facilities. For unsteady flow, HEC RAS solves the one-dimensional Saint Venant equation using an implicit, finite difference method. HEC-RAS has common applications in floodplain management and flood insurance studies. The main window of the HEC-RAS model is shown in Figure 7.7. In this investigation, the virgin mode of the HEC-RAS model was adapted, in which rainfall is only taken as input without any external sources of flood triggering. Besides the exclusive input to this model, rainfall was classified into four modes: normal, moderate, and heavy. The details of that sorting are shown in Figure 7.8.

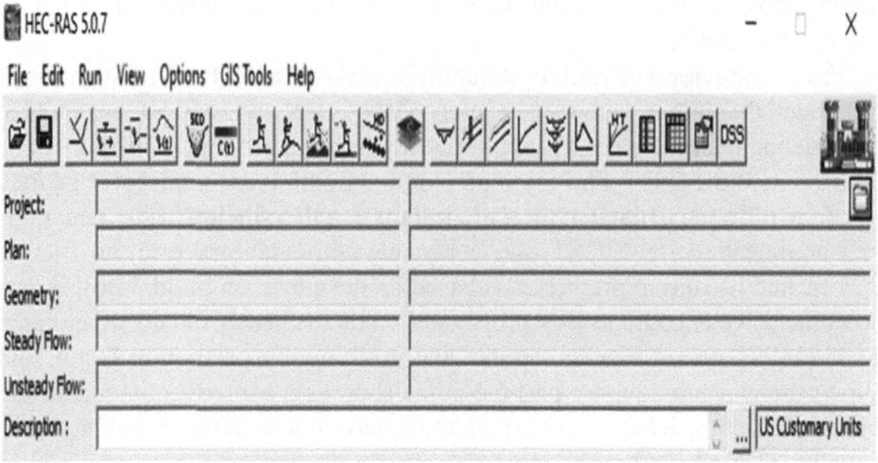

FIGURE 7.7 The main window of HEC-RAS model.

Intensity of Rainfall	
V L	Very Light Rainfall (0.1 to 2.4 mm)
L	Light rainfall (2.5-15.5 mm)
M	Moderate (15.6-64.4 mm)
H	Heavy Rainfall (64.5-115.5 mm)
VH	Very Heavy Rainfall (115.6-204.4 mm)
X H	Extremely Heavy Rainfall (>204.4mm)

FIGURE 7.8 Intensity sorting of Rainfall.

7.6.2 BASIC FLOOD MODELLING EQUATIONS

The Navier-Stokes equations are the fundamental equations used in fluid dynamics. Hydraulic models rely on two main governing equations of physics: the equation of continuity and the conservation of momentum. These equations are employed in hydraulic models to ensure simplicity, accuracy, and stability.

According to the continuity of mass:

$$\frac{\partial Q}{\partial x}+\frac{\partial A}{\partial x}-q=0 \tag{3}$$

The momentum equation from Newton's second law is:

$$\frac{1}{A}\frac{\partial Q}{\partial t}+\frac{1}{A}\frac{\partial}{\partial x}\left(\frac{Q^2}{A}\right)+g\frac{\partial y}{\partial x}-g\left(S_0-S_f\right)=0 \tag{4}$$

Equations (3) and (4) together are referred to as the Saint Venant equation. This equation has various applications and is widely used to predict the arrival time and magnitude of floods. The solution of the Saint Venant equations provides the variation of discharge over time, which is used for flood forecasting.

7.6.3 MODELLING INPUTS AND OUTPUTS

The inputs and outputs of modeling concerned with tsunamis and floods are mentioned in Figure 7.9.

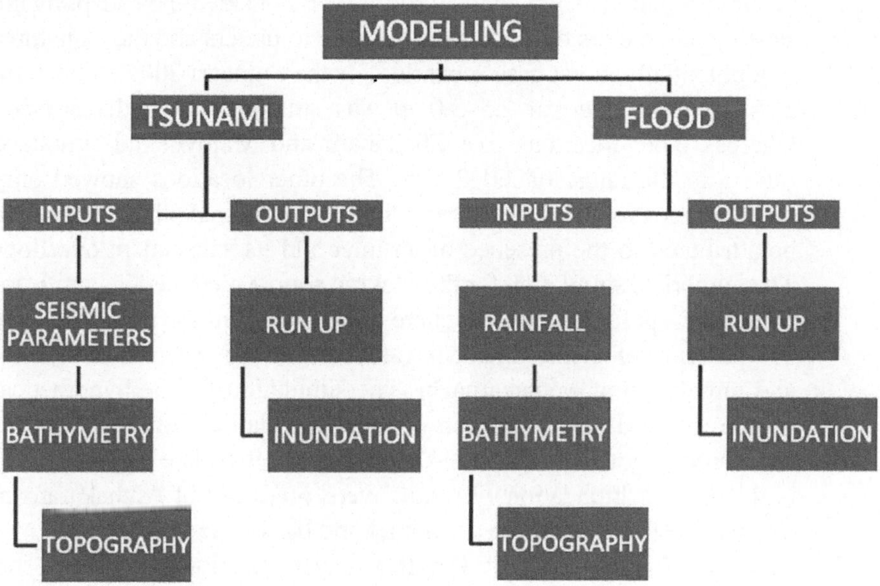

FIGURE 7.9 Modelling inputs and outputs.

For both the bathymetry and terrain topography of these models, they play a prominent role in shaping inundation and run-up characteristics along the coast and banks.

7.7 RESULT AND DISCUSSION

The simulation results for tsunami and floods are mentioned in subsections.

7.7.1 RESULTS OF TSUNAMI MODELLING

The results of tsunami inundation and run-up modeling fall under two sectors pertaining to Ernakulam district (Figure 7.10 and Table 7.2):

1. **Vayalar–Manasseri:** The simulation was carried out for three different sources – (i) Sumatra 2004; (ii) Makran 1945; and (iii) hypothetical source. The earthquake intensity of the Sumatra earthquake was around 9.3, while the Makran earthquake in 1945 was around 8.1. When considering the state of Kerala, the Sumatra source is located on the eastern side and Makran on the western side. Makran is close to Kerala in terms of proximity. Although its intensity was lower, in this investigation, a hypothetical source was considered by adapting the seismic parameters of Sumatra in Makran to understand the signatures of a potentially worst-case scenario. For the Sumatra 2004 simulation, a maximum run-up of 2.5–3.0 m was simulated at Anthakarnazhi, whereas other locations like Chellanam and Maravukkad simulated run-up in the range of 2.0–2.5 m. The other locations showed only feeble inundation, and the increase in inundation in Anthakarnazhi can be attributed to the presence of an inlet and its subsequent overflow. The numerical simulation for the Makran source predicted insignificant values, except for Manasseri, where it showed a run-up value of 2.12 m. The presence of an inlet and backwater contributed to significant run-up and inundation at Anthakarnazhi. The simulation for the hypothetical source showed increased run-up and inundation for all locations compared to both Sumatra 2004 and Makran 1945. The highest run-up (2.44 m) and inundation (2.4 km) were predicted at Anthakaranzhi, mainly due to the presence of an inlet and backwaters.
2. **Manasseri–Munambam:** For this sector, the location Manasseri showed a high run-up for the Sumatra simulation, ranging from 2.0 to 2.5 m. However, other locations like Vypin and Puthuvypin showed run-up in the range of 1.5 to 2.0 m. The model predicted an inundation of 345 m at Vypin, 853 m at Puthuvypin, and 670 m at Njarackal. The numerical simulation for Makran 1945 showed a consistent run-up pattern ranging from 2.0 to 2.5 m along the entire

coast. The inundation along this stretch for the Makran simulation was also relatively high, with Vypin and Puthuvypin experiencing 722 m and 1,060 m of inundation, respectively.

The model simulated increased run-up and inundation along the entire stretch for the hypothetical source, comparing it with that of Makran 1945 and even Sumatra 2004. A high inundation value of 2.76 m was observed at Manasseri, which is considerably higher than the Sumatra 2004 case (Figure 7.10 and Table 7.2).

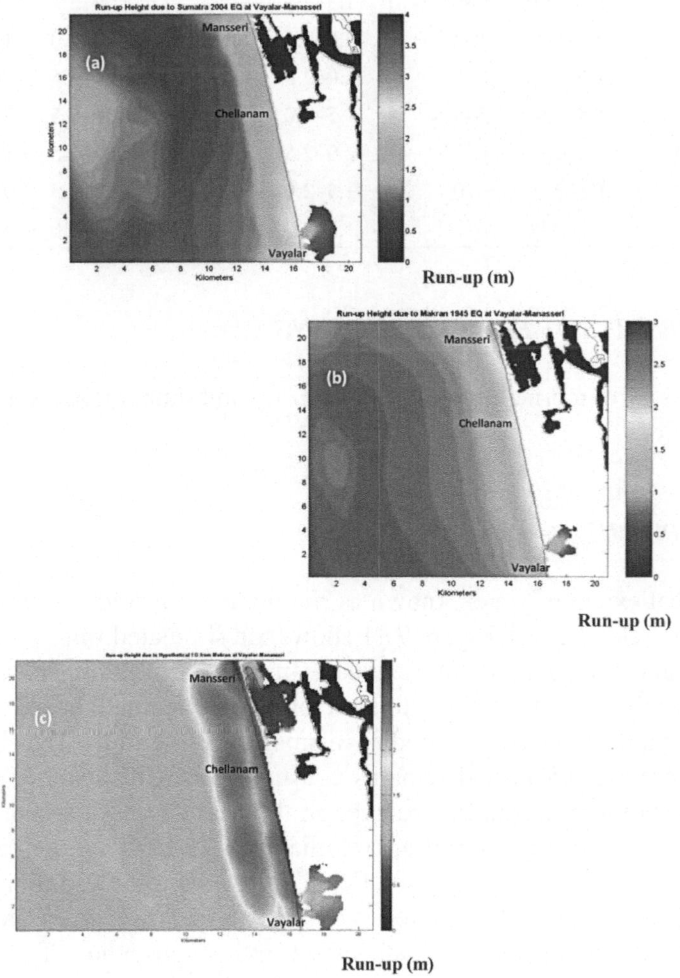

FIGURE 7.10 Model simulation output for the Vayalar-Manasseri coast of Ernakulam district includes: (a) Sumatra 2004; (b) Makran 1945; and (c) hypothetical worst case.

TABLE 7.2 Run-Up and Inundation Along the Coastal Stretch of Ernakulam District for Three Different Cases of Tsunami Generation

Location	Sumatra 2004		Makran 1945		Hypothetical	
	Run-Up (m)	Inundation (m)	Run-Up (m)	Inundation (m)	Run-Up (m)	Inundation (m)
Kadapuram	2.0–2.5	Nil	1.0–1.5	Nil	2.0–2.5	Nil
Anthakarnazhi	2.5–3.0	2,000	1.5–2.0	1,150	2.0–2.5	2,400
Chellanam	2.0–2.5	Nil	1.5–2.0	Nil	2.5–3.0	45
Maravukad	2.0–2.5	30	1.5–2.0	32	2.0–2.5	47
Kannamali	1.5–2.0	186	1.5–2.0	94	2.0–2.5	2,370
Manasseri	2.0–2.5	1,102	2.0–2.5	1,102	2.5–3.0	1,500
Fort Cochin	1.5–2.0	Nil	2.0–2.5	Nil	2.0–2.5	80
Vypin	1.5–2.0	345	1.5–2.0	722	2.0–2.5	722
Puthuvypin	1.5–2.0	853	2.0–2.5	1,060	2.0–2.5	1,060
Njarackal	1.5–2.0	670	1.5–2.0	925	2.0–2.5	925
Edvanakkad	1.5–2.0	Nil	1.5–2.0	Nil	1.5–2.0	Nil

7.7.2 RESULTS OF FLOOD MODELLING

The results of numerical modelling of run-up and inundation are mentioned in subsections.

7.7.2.1 RUN UP

The vertical extent of water, known as run-up, is measured with reference to MSL (Mean Sea Level). Figure 7.11 shows the simulated run-up for all four scenarios and compares it with field data. In the case of normal rainfall, there is consistently no run-up along this sector from Cheranalloor to Aluva. For moderate rainfall, the model predicts a small increase in the run-up pattern, which is around 0.5 m and remains constant from Cheranalloor to Aluva. A significant increase can be observed in the heavy rainfall scenario, where the model predicts a run-up of approximately 1.25 m along the banks. It is important to note that the observed field data shows some variation from the simulated field data. This variation is due to external discharges such as dam discharge, tidal reflection, backwater effect, and various others. The simulated run-up depends solely on the rainfall data because it is a virgin model, and no other external sources are incorporated. Locations like Marampally,

Chalakkal, Thekkumbhagam, Kuttamsserry, Thottumbhagam, and Aluva simulated a run-up height of 1.25 for heavy rainfall. The observed values for these locations are slightly higher as external factors also influenced the discharge during the 2018 Kerala floods. The simulation for hypothetical rainfall, including extreme rainfall events, predicted a drastic increase in run-up.

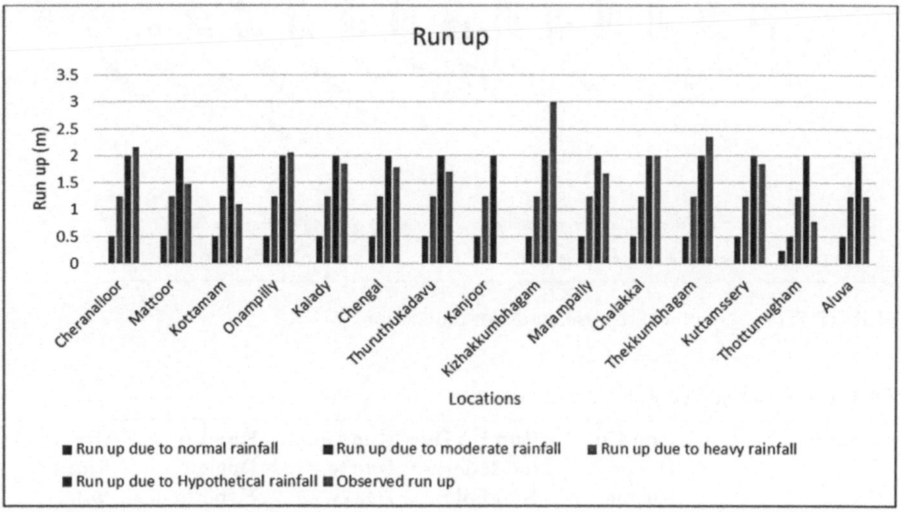

FIGURE 7.11 Graphical representation of observed run up and simulated run up values.

Based on the simulation results, it can be concluded that the terrains along Central Kerala will not be inundated due to normal and moderate rainfall events. However, heavy rainfall can cause significant inundation. In the case of extreme rainfall events, such as the hypothetical source of modeling run-up, the inundation can increase drastically. Additionally, the higher run-ups in certain locations can be attributed to the prevailing terrain elevation in that sector.

7.7.2.2 INUNDATION

Inundation shows the horizontal extent of water that has entered the bank and is measured with reference to the shoreline of the river. The model predicted inundations for all four rainfall scenarios, and the results are summarized in Figure 7.12 and Table 7.3.

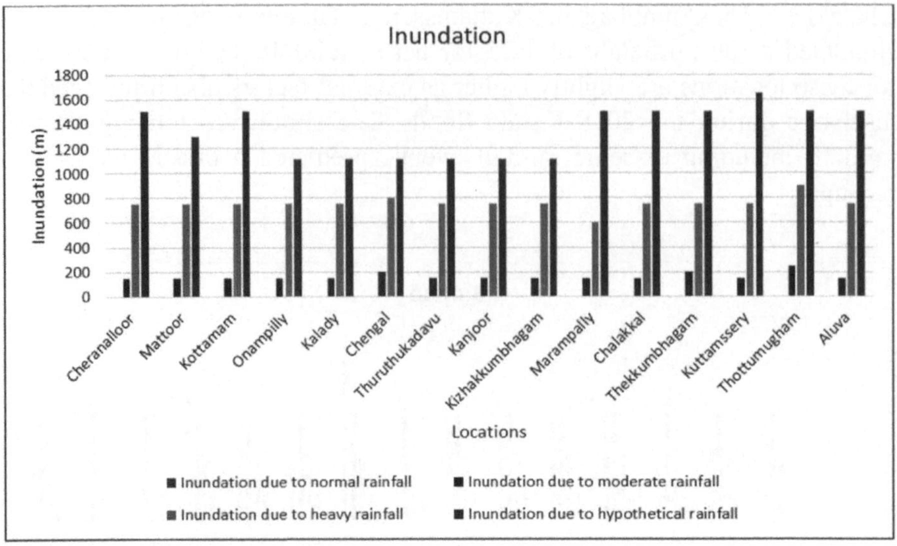

FIGURE 7.12 Graphical representation of inundation.

TABLE 7.3 Observed Run Up and Simulated Run Up Values

Locations	Run Up Due to Normal Rainfall (m)	Run Up Due to Moderate Rainfall (m)	Run Up Due to Heavy Rainfall (m)	Run Up Due to Hypothetical Rainfall (m)	Observed Run Up (m)
Cheranalloor	0	0.5	1.25	2	2.16
Mattoor	0	0.5	1.25	2	1.48
Kottamam	0	0.5	1.25	2	1.1
Onampilly	0	0.5	1.25	2	2.07
Kalady	0	0.5	1.25	2	1.85
Chengal	0	0.5	1.25	2	1.79
Thuruthukadavu	0	0.5	1.25	2	1.7
Kanjoor	0	0.5	1.25	2	0
Kizhakkumbhagam	0	0.5	1.25	2	3
Marampally	0	0.5	1.25	2	1.67
Chalakkal	0	0.5	1.25	2	2
Thekkumbhagam	0	0.5	1.25	2	2.35
Kuttamssery	0	0.5	1.25	2	1.85
Thottumugham	0.25	0.5	1.25	2	0.78
Aluva	0	0.5	1.25	2	1.25

The model predicted no inundations for normal and moderate ranges of rainfall. However, for heavy rainfall and extreme rainfall events, significant and alarming inundations were predicted, respectively. This pattern was uniformly observed in the locations from Cheranalloor to Aluva for both of these sources. The local elevation, with reference to sea level, also contributed to the inundation, which reached approximately 0.5 meters. Unfortunately, the inundation results could not be validated with field data due to its unavailability (Table 7.4).

TABLE 7.4 Simulated Inundation Along the Stretch

Locations	Inundation Due to Normal Rainfall (m)	Inundation Due to Moderate Rainfall (m)	Inundation Due to Heavy Rainfall (m)	Inundation Due to Hypothetical Rainfall (m)
Cheranalloor	0	150	750	1,500
Mattoor	0	150	750	1,300
Kottamam	0	150	750	1,500
Onampilly	0	150	750	1,110
Kalady	0	150	750	1,110
Chengal	0	200	800	1,110
Thuruthukadavu	0	150	750	1,110
Kanjoor	0	150	750	1,110
Kizhakkumbhagam	0	150	750	1,110
Marampally	0	150	600	1,500
Chalakkal	0	150	750	1,500
Thekkumbhagam	0	200	750	1,500
Kuttamssery	0	150	750	1,650
Thottumugham	0	250	900	1,500
Aluva	0	150	750	1,500

7.8 CONCLUSIONS

The state of Kerala, situated in the southern peninsula of India, has been prone to various natural disasters. Tsunamis, cyclones, floods, and landslides have exposed the vulnerability of Kerala. In response to each disaster, research and development efforts have been intensified to establish a warning system that provides advisories for mitigation and rehabilitation. Numerical modeling plays a crucial role in these warning systems. In this study, numerical modeling was conducted to assess the impact of vulnerability caused by tsunamis and cyclones in a selected location of Kerala. Additionally, the physics and modeling concepts behind these disasters are also discussed.

ACKNOWLEDGMENTS

The authors express their deep gratitude to the Principal and Staff of St. Thomas College, Kozhencherry for their invaluable support. Special thanks are extended to Dr. Ann Rose Abraham, Assistant Professor of Physics at Sacred Heart College Thevara, Ernakulam, for her significant contribution to the realization of this investigation.

KEYWORDS

- **flood**
- **flood modelling equations**
- **hypothetical rainfall**
- **inundation**
- **mitigation**
- **modelling**
- **rehabilitation**
- **run-up**
- **tsunami**

REFERENCES

1. Praveen, S. S., (2012). PhD Thesis, *Numerical Modeling of Tsunami Propagation in the South East Arabian Sea (SEAS) and Inundation Along the Kerala Coast.* Cochin University of Science and Technology, Centre for Earth Science Studies.
2. Goto, C., & Ogawa, Y., (1992). *Numerical Method of Tsunami Simulation with the Leapfrog Scheme.* Translated for the TIME Project by N. Shuto, Disaster Control Res. Cent., Faculty of Engg. Tohoku Univ. Japan.
3. Mercado, A., & McCann, W., (1998). Numerical simulation of the 1918 Puerto Rico tsunami. *Natural Hazards, 18*, 57–76.

4. Yalciner, A. C., Alpar, B., Altinok, Y., Ozbay, I., & Imamura, F., (2002). Tsunamis in the sea of Marmara: Historical documents for the past, models for future. *Mar. Geol., 190*, 445–463.
5. Yalciner, A. C., Dogan, P., Sukru, E., Gegar, S., Presteya, G. S., Hidayat, R., & McAdoo, B., (2005). Indian Ocean tsunami field survey, north of Sumatra Island, UNESCO, http://ioc.unseco.org/iosurveys/Indonesia/yalciner/yalciner.htm (accessed on 16 January 2023).
6. Zahibo, N., Pelinovsky, E., Yalc, A., Kurkin, A., Koselkov, A., & Zaitsev, A., (2003). The 1867 virgin Island tsunami: Observations and modeling. *Oceanologica Acta, 26*, 609–621.
7. Fujima, K., Masamura, K., & Goto, C., (2002). Theoretical examinations of long-wave damping by sea bottom friction. *Coast. Engg. J., 44*, 217–245.
8. Kajiura, K., (1963). The leading wave of a tsunami. *Bull. Earthq. Res. Inst., 41*, 535–571.
9. Shuto, N., & Goto, C., (1978). Numerical simulation of tsunami run-up. *Coast. Engg. Japan, 21*, 13–20.
10. Takeda, H., (1984). Numerical simulation of run-up by variable transformation. *Oceanogra. J. Japan,40*, 271–278.
11. Hibbard, S., & Peregrine, Y., (1979). Surf and run-up on a beach: A uniform bore. *J. Fluid Mech., 95*, 323–345.
12. Aida, I., (1977). Tsunamis accompanied by landslides, Kaiyo-Kagaku. *J. Mar. Sci. Japan, 9*, 103–110.
13. Houston, J. R., & Butler, H. L., (1979). *A Numerical Model for Tsunami Inundation.* U.S. Army Corps of Engineers, Hydraulic Laboratory, Waterways Experiment Station, Viksburg, Mississippi, *Tech. Rep.* HL-79-2.
14. Goto, C., & Shuto, N., (1983). Effects of large obstacles on tsunami inundations. In: Iida, K., & Iwasaki, T., (eds.), *Tsunamis: Their Science and Engineering* (pp. 521–525). Tokyo.
15. Uda, T., (1988). Numerical simulation and experiment on tsunami run-up. *Coast. Engg. Japan, 31*, 87–104.
16. Kajiura, K., (1970). Tsunami source, energy, and the directivity of wave radiation. *Bull. Earthq. Res. Inst., 48*, 835–869.
17. Mansinha, L., & Smylie, D. E., (1971). The displacement field of inclined faults. *Bull. Seism. Soc. America, 61*, 1433–1440.
18. Okada, Y., (1985). Surface deformation due to shear and tensile faults in a half-space. *Bull. Seism. Soc. America, 75*, 1135–1154.
19. Imamura, F., (1996). Review of tsunami simulation with a finite difference method. In: Yeh, H., Liu, P., Synolakis, C., & Hackensack, N. J., (eds.), L*ong Wave Run-Up Models* (pp. 25– 42). World Scientific Publishing Co.
20. Praveen, S. S., & Sajith, B. S., (2018). Numerical modelling of floods in Kerala-a research and developmental perspective. *Rational Discourse, 24*(1).
21. Yalciner, A. C., Pelinovsky, E., Talipova, T., Kurkin, A., Kozelkov, A., & Zaitsev, A., (2004). Tsunamis in the black sea: Comparison of the historical, instrumental, and numerical data. *J. Geophys. Res., 109*, 2003–2113.
22. Goto, C., Ogawa, Y., Shuto, N., & Imamura, F., (1997). *Numerical Method of Tsunami Simulation with the Leap-Frog Scheme (IUGG/IOC Time Project).* IOC Manual, UNESCO, No. 35.

Index

C

D

E

F

N

T